Synthesis Lectures on Digital Circuits & Systems

Series Editor

Mitchell A. Thornton, Southern Methodist University, Dallas, USA

This series includes titles of interest to students, professionals, and researchers in the area of design and analysis of digital circuits and systems. Each Lecture is self-contained and focuses on the background information required to understand the subject matter and practical case studies that illustrate applications. The format of a Lecture is structured such that each will be devoted to a specific topic in digital circuits and systems rather than a larger overview of several topics such as that found in a comprehensive handbook. The Lectures cover both well-established areas as well as newly developed or emerging material in digital circuits and systems design and analysis.

Zarin Tasnim Sandhie · Farid Uddin Ahmed ·
Masud H. Chowdhury

Beyond Binary Memory Circuits

Multiple-Valued Logic

Zarin Tasnim Sandhie
Electrical and Computer Engineering
University of Missouri–Kansas City
Kansas City, MO, USA

Farid Uddin Ahmed
Electrical and Computer Engineering
University of Missouri–Kansas City
Kansas City, MO, USA

Masud H. Chowdhury
Electrical and Computer Engineering
University of Missouri–Kansas City
Kansas City, MO, USA

ISSN 1932-3166   ISSN 1932-3174 (electronic)
Synthesis Lectures on Digital Circuits & Systems
ISBN 978-3-031-16197-1   ISBN 978-3-031-16195-7 (eBook)
https://doi.org/10.1007/978-3-031-16195-7

This Springer imprint is published by the registered company Springer Nature Switzerland AG
The registered company address is: Gewerbestrasse 11, 6330 Cham, Switzerland

Dedicated to our beloved parents, spouses, children, and family members.

Preface

About the Subject

Computing technologies are currently based on the binary logic/number system, which is dependent on the simple on and off switching mechanism of the prevailing transistors. With the exponential increase in data processing and storage needs, there is a strong push to move to a higher radix logic/number system that can eradicate or lessen many limitations of the binary system. Anticipated saturation of Moore's law and the necessity to increase information density and processing speed in the future micro and nanoelectronic circuits and systems provide a strong background and motivation for the beyond binary logic system.

There are lots of research proving that the ternary logic system is the most efficient and cost-effective than other logic system. In a ternary system (base 3 or more), each logic bit can have three possible values leading to significantly higher information density with smaller logic gates and reduced circuit complexity. As a result, the energy consumption, area, and circuit overheads, and other costs for each bit of information would decrease in the ternary system. The computational complexity decreases manyfold in a ternary system, which in turn enhances the power and area efficiency of the whole system. The existing technologies like MOSFET, FinFET, and FDSOI are not flexible enough to hold three distinguished voltage levels with proper noise margin. Hence, the concept of using different emerging technologies comes into view.

The impact of my multi-valued logic (MVL)-based works can be tremendous. The world as we know is mostly binary technology based. For decades, researchers have been trying to move toward higher-base logic systems, preferably ternary or quaternary logic systems. In few communication and memory sectors, a higher-base system has been proved to be of great advantage and is used industrially. Replacing a small part (especially any memory circuit) of any digital device, the consequence can be enormous. It can enhance the information holding capacity of a device to a great extent. Memory circuits and systems are the indispensable parts of computing and most of the other micro and nanoelectronic applications. The efficiency of memories is mainly dependent on the density of information. One significant way to increase the information density is to hold more data in a single bit, which can be achieved through the introduction of MVL technologies in memory applications. There are many individual and collaborative projects currently underway to explore beyond binary logic and memory technologies. Researchers are offering different design techniques to utilize these MVL technologies in different types of memory circuits like flash memory, content-addressable memory, etc.

About This Book

This book has been written to address a critical need of the digital circuit community involved in conducting research on ternary/quaternary logic circuits or emerging technologies. There has been a lot of approaches to develop a multi-valued logic system as well as to develop newer circuit design to implement those logic. For a long time, researchers have been trying to figure out which radix works better in terms of circuit complexity and cost-effectiveness, and which design approach or technology implementation is better for developing MVL circuits.

A handful of books are available on the topics of arithmetic representations and synthesis techniques of MVL. However, there is no dedicated book that combines all the design implementation and technology implementation of MVL memory circuits. The authors of this book intend to provide a single and comprehensive resource on MVL-based memory technologies for a wide range of readers, including the students, trainees, and researchers in the academic and industrial communities. This book covers all the significant works done in this emerging area of research. It features essential design concepts for different types of multiple-valued memory circuits, such as sequential circuit, static random-access memory (SRAM), dynamic random-access memory (DRAM), ternary content-addressable memory (TCAM), flash memory, etc.

This book will give the reader an in-depth understanding of how the multi-valued logic works, why it is crucial, and its historical importance. After that, it will give the reader a basic understanding of the past and recent development in the field of MVL-based memory circuits. Also, this book will help the reader to design more efficient memory circuits using different prevailing and emerging technologies. In addition to giving a basic concept of the multi-valued logic system, it also gives a complete survey of the different designs presented in different pieces of literature on multi-valued logic-based memory circuits.

Organization of the Book

Chapter 1 outlines the basic definition and computational advantage of multi-valued logic (MVL). Also, the historical background and different scopes of the MVL circuit are introduced in this chapter. The challenges and future direction of MVL are also included here.

Chapter 2 provides the signal representation, basic mathematical operations, and a few important synthesis techniques of MVL.

Chapter 3 discusses the different technologies tried for the implementation of MVL circuits. Here a brief discussion about the design methodology of obtaining multiple voltage levels using the available technologies and a thorough comparison between those methods is given.

Chapter 4 contains the basic principles of various MVL sequential circuits like latch, flip-flop, etc. It also gives a comparison between different available MVL sequential circuit designs.

Chapter 5 discusses the design principle of ternary static random-access memory (SRAM), dynamic random-access memory (DRAM), and multi-level DRAM (MLDRAM).

Chapter 6 presents the design methodology and analysis of MVL NAND and NOR flash memory.

Chapter 7 is dedicated to the discussion of ternary content-addressable memory (TCAM). Here, the different approach for designing TCAM circuits as well as their comparison is given.

Kansas City, USA

Zarin Tasnim Sandhie
Farid Uddin Ahmed
Masud H. Chowdhury

Acknowledgements

We are thankful to all who contributed one way or another in the completion of this book. First and foremost, thanks to almighty God for His showers of blessings in our lives.

We are deeply thankful to all the professors at University of Missouri-Kansas City for all their encouragement and guidelines, specially to Professor and Chair, Dr. Ghulam Chaudhry. We would like to thank our group members (Abdul Hamid Bin Yousuf, Abdullah G. Alharbi, Athiya Nizam, Azzedin EsSakhi, Emeshaw Ashenafi, Hemanshu Shishupal, Liaquat Ali, Mahmood Uddin Mohammed, Marouf Khan, Moqbull Hossen, Munem Hossain) at the Micro and Nano Electronics Laboratory in the Department of Computer Science Electrical Engineering (CSEE), School of Computing and Engineering (SCE), University of Missouri–Kansas City (UMKC), for their support.

Most importantly, we are extremely thankful to our parents, for their love, prayers, caring and sacrifices for our education and preparing us for our future. We are eternally grateful to our spouses and children for their love, understanding and continuous support.

Contents

About the Authors

Zarin Tasnim Sandhie received the B.Sc. degree in Electrical and Electronic Engineering from Bangladesh University of Engineering and Technology (BUET), Dhaka, Bangladesh in 2015. She received her M.S. and Ph.D. degrees in Electrical and Computer Engineering from University of Missouri-Kansas City in 2020 and 2021, respectively. She authored ~10 research publications during her M.S. and Ph.D. researches. Her research interests include multiple-valued logic (MVL), ternary logic, and graphene nanoribbon field-effect transistor (GNRFET). Zarin was awarded the UMKC School of Graduate Studies Research Grant in 2018. She got the 1st prize in UMKC hack-a-roo 2018. She completed her internship as a graduate intern at Intel in the Atom CPU Design Team. Currently, she is working as SoC Design Engineer in the IPG group in Intel Corporation.

Farid Uddin Ahmed received the B.Sc. degree in Electrical and Electronic Engineering from Bangladesh University of Engineering and Technology (BUET), Dhaka, Bangladesh in 2015. He completed his M.S. and Ph.D. from the Computer Science and Electrical Engineering Department of University of Missouri-Kansas City (UMKC), USA. He authored/co-authored 11 publications as a Ph.D. student. Farid was awarded the UMKC School of Graduate Studies Research Grant in 2017. Currently, he is working as SoC Design Engineer in the Standard Cell library Design Team at Intel.

His research interests include on-chip power management integrated circuit (PMIC), design automation of analog circuit, and multi-valued logic (MVL).

Masud H. Chowdhury received the B.S. degree in Electrical and Electronic Engineering from Bangladesh University of Engineering and Technology (BUET), Dhaka, in 1998 and the Ph.D. degree in Computer Engineering from Northwestern University, Evanston, Illinois, USA, in 2004. Currently, he is the Interim Department Chair and Associate Dean of the School of Computing and Engineering at the University of Missouri–Kansas City (UMKC). He has published more than 150 articles in various journals and conferences in his fields of research, which include (i) high-performance issues of deep submicron and nanoscale integrated circuits and (ii) emerging 2D nanomaterials and ferroelectric material based devices and circuits for computing, memory, sensing, and energy applications. He is the co-founder and director of Center for Interdisciplinary Nano Technology Research (CINTR) at UMKC. Dr. Chowdhury has served as the Chair of IEEE VLSI Systems and Applications Technical Committee from 2014 to 2016. He is also serving as the Associate Editor of (i) *IEEE Transactions on VLSI Systems*, (ii) *Microelectronics Journal* of Elsevier, (iii) *IEEE Transactions on Circuits and Systems II* (TCAS II), and (iv) *Journal of Circuits, Systems, and Signal Processing*, Springer. He has been serving the professional communities in many other capacities for more than a decade.

Abbreviations

ALU	Arithmetic logic unit
BJT	Bipolar junction transistor
BL	Bit line/Bitline
BLIF	Berkeley Logic Interchange Format
BOX	Buried insulating oxide
CAM	Content-addressable memory
CNT	Carbon nanotube
CNTFET	Carbon nanotube field-effect transistor
DBQW	Double-barrier quantum well
DFF	D flip-flop
DFFF	D flip-flap-flop
DIBL	Drain-induced barrier lowering
DRAM	Dynamic random-access memory
ECC	Error-correcting codes
EEPROM	Electronically erasable programmable read-only memory
FDSOI	Fully depleted silicon on insulator
FF	Flip-flop
FGMOS	Floating gate metal oxide semiconductor
FinFET	Fin field-effect transistor
GNR	Graphene nanoribbon
GNRFET	Graphene nanoribbon field-effect transistor
GTI	General ternary inverter
HBT	Heterojunction bipolar transistor
HDL	Hardware description language
HEMT	High electron mobility transistors
ISI	Intersymbol interference
LSB	Least significant bit
MIFG	Multiple input floating gate
MIT	Metal-insulator transition
MLC	Multi-level cell

MLDRAM	Multi-level dynamic random-access memory
MOBILE	Monostable-to-bistable transition logic elements
MODFET	Modulation-doped field-effect transistor
MOSFET	Metal oxide semiconductor field-effect transistor
MSB	Most significant bit
MTJ	Magnetic tunnel junction
MVL	Multiple-valued logic
MVSIS	Multi-valued sequential interactive synthesis
NDR	Negative-differential resistance
Neuron-MOS	Neuron metal oxide semiconductor
NMOS	N-type metal oxide semiconductor
NTI	Negative ternary inverter
NVRWM	Non-volatile read–write memory
PAM	Pulse amplitude modulation
PDSOI	Partially depleted silicon on insulator
PISO	Parallel-in-serial-out
PMOS	P-type metal oxide semiconductor
PTI	Positive ternary inverter
QAM	Quadrature amplitude modulation
QDGFET	Quantum dot gate field-effect transistor
QLC	Quad level cell
QPSK	Quadrature phase shift keying
ROM	Read-only memory
RTD	Resonant tunneling diode
RWM	Read–write memory
SCE	Short-channel effects
SET	Single-electron transistor
SIPO	Serial-in-parallel-out
SIS	Sequential interactive synthesis
SISO	Serial-in-serial-out
SOI	Silicon on insulator
SOT	Spin-orbit torque
SRAM	Static random-access memory
STI	Standard ternary inverter
STT	Spin transfer torque
TCAM	Ternary content-addressable memory
TLC	Triple-level cell
TMR	Tunneling magnetoresistance
UTBB-SOI	Ultra-thin-body buried oxide SOI
VCMA	Voltage-controlled magnetic anisotropy
WL	Word line/Wordline

List of Figures

List of Tables

Background and Future of Multiple Valued Logic

1.1 Introduction

In 1965, Gordon Moore, the co-founder of Intel, predicted that the transistor count in a fabricated chip would double every two years. Since then, the continuous increase of transistor density and chip fabrication efficiency resulted in the microprocessor performance doubling every couple of years [1]. The prediction, which became known as Moore's law, has guided the phenomenal growth of the integrated circuits and computing industry for the last six decades. The continuous increase of integration density has been possible due to the shrinkage of transistor sizes leading to extremely compact, energy-efficient, and powerful processors. However, as the transistor dimensions are approaching their physical and material limits, adopting new and subsequent generations of smaller and faster transistors is becoming increasingly challenging. Several well-known semiconductor industries are creating chips containing transistors of 7 nm technology, and further shrinking and mass production of smaller transistors are becoming more complex and expensive [1, 2]. The graphical illustration of Fig. 1.1 [3] represents the number of transistors count in Intel microprocessors in the last six decades.

The limits of technology scaling and the anticipated saturation of Moore's Law are imposing fundamental technological and physical roadblocks for binary logic-based applications built on silicon and CMOS platforms. As a result, industry and academia have been exploring several groundbreaking paths forward: (1) introduce post-silicon technologies to fabricate chips with sub-nanometer dimensions, (2) utilize non-CMOS technologies to implement binary logic, and (3) move to beyond-binary (base 3 or higher) logic system.

Though the concept of Multiple-Valued Logic has been known since the time of Aristotle, binary or classical logic is the one which has been studied and applied in the real world for the longest period of time. Since the last century, Multiple-Valued-Logic (MVL)

© The Author(s), under exclusive license to Springer Nature Switzerland AG 2022
Z. T. Sandhie et al., *Beyond Binary Memory Circuits*, Synthesis Lectures
on Digital Circuits & Systems, https://doi.org/10.1007/978-3-031-16195-7_1

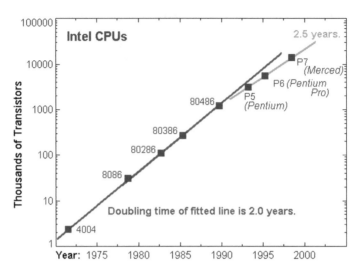

Fig. 1.1 Moore's law and transistor count per chip with respect to time [3]

started gaining attention after Łukasiewicz in 1920 attempted to create a many-valued logic system that included a third value "possible" [4]. In 1921, Emil L. Post presented the interpretation of additional truth degrees with $n \geq 2$, where n is the number of truth values [5]. After that, few other propositions contributed to the conceptualization of many-valued logic. However, none of them were completely applied in the real world as binary logic. The ON–OFF switching mechanism of the transistors contributed to the widespread use of binary logic in today's computing system. The binary or Boolean numeric system has two discrete logic levels as 0 (False) and 1 (True). In the case of MVL, more than two logic levels are required to define the system. Depending on the number of levels, MVL can be ternary (three levels), quaternary (four levels), and so on. The potential applications of MVL are wide-ranging and can be classified into two primary approaches. The first approach is to utilize the MVL system as a background for solving binary problems more effectively. And the second approach is to design newer devices and circuits that can work on a multi-valued platform. A lot of work is being done using prevailing and emerging technologies to implement the MVL circuits. MVL holds many advantages over binary logic. It permits a single bit to hold more information than a binary bit, resulting in reduced wiring complexity, lower energy consumption, smaller area, and circuit overheads, and lower costs for each bit of data [6]. Researchers have been trying to implement both arithmetic and memory circuits (two vital components of any computation system) using MVL technologies during the last few decades. This book presents an overview of beyond-binary or multiple-valued logic technologies for memory applications.

1.2   What is Multiple-Valued Logic

An essential principle of the Classical or Aristotle's logic is the *principle of bivalence*, which states that there are only two possible values for any proposition (true and false). Whether this fundamental principle remains true in every situation has always been a topic of debate both in logic and philosophy [7]. In a logical system, a many-valued logic or multiple-valued logic is the branch of propositional calculus with more than two truth values. For Multiple-Valued Logic (MVL), classical two-valued logic is extended to an *n*-valued system where *n* can be any number greater than two. We have used many-valued logic, multi-valued logic, and multiple-valued logic synonymously throughout this book.

The most common and well-accepted forms of many-valued logic available in the literature are (i) the three-valued logic (Łukasiewicz's and Kleene) that denotes the three values as "true," "false," and "unknown"; (ii) the finite-valued logic where there are finite number of distinct values; and (iii) the infinite-valued logic where there are an infinite number of distinct values, such as fuzzy logic [8].

1.3   Computational Advantages of MVL System

Decimal (base 10) and binary (base 2) number systems are the two most widely used number systems in real-life calculations and digital applications, respectively. However, neither base 2 nor base 10 system is the most efficient to work with from a computation efficiency point of view [9]. Theoretically, it has been proven that the base-3 number system would be the most efficient under an ideal situation.

In the case of any numerical system, the number of digits required to express a specific number is inversely proportional to the radix number. Let, the number range is denoted by N, the radix is denoted by R, and d is the required number of digits (d is rounded to the subsequent integer value). The relationship can be expressed by (1.1). Table 1.1 gives an overview of the information handling capability (calculated using Eq. (1.1)) of binary, ternary, and quaternary logic systems.

Table 1.1 Table showing the density of information by 2, 3, and 4 valued logics

Highest possible data or the range of number (N) that can be contained by	Binary (R = 2)	Ternary (R = 3)	Quaternary (R = 4)
1 bit	$2^1 = 2$	$3^1 = 3$	$4^1 = 4$
2 bit	$2^2 = 4$	$3^2 = 9$	$4^2 = 16$
3 bit	$2^3 = 8$	$3^3 = 27$	$4^3 = 64$
4 bit	$2^4 = 16$	$3^4 = 81$	$4^4 = 256$

$$N = R^d \tag{1.1}$$

Taking logarithm for both sides of (1.1), we get:

$$\Rightarrow \log N = d \times \log R \tag{1.2}$$

$$\Rightarrow d = \frac{\log N}{\log R}$$

The cost or the complexity of the system hardware C is proportional to the digit capacity $(R \times d)$. Then from (1.2), it can be obtained that,

$$C = k(R \times d) = k\left[R\frac{\log N}{\log R}\right] \tag{1.3}$$

$$\Rightarrow C = k\left[R\frac{\log N}{\log R}\right]$$

Here, k is the constant of proportionality. By differentiating (1.2) with respect to R, it can be proved that for the minimum value of cost C, the radix number R is equal to $e = 2.72$. If we round the value of e to its nearest integer, we can say that for $R = 3$ or a ternary system, the system's cost or complexity will be minimum.

If we consider a system where the cost of system C is independent of the system radix R, then (1.3) can be written as (1.4).

$$C = kd = k\left[\frac{\log N}{\log R}\right] \tag{1.4}$$

From (1.4), it is apparent that the cost or complexity of circuit C is inversely proportional to the radix R. In that case, a ternary system is more cost-efficient than a binary system [10]. The findings of the above calculations are summarized in Table 1.2. The results of the table do not include the cost of manufacturing MVL circuits or the nonbinary calculations for the numerical data, which is largely dependent on the technology involved and the nonbinary circuit designs. Besides these, there are some notions of quaternary logic being the most-efficient logic system in few literature [11].

In addition to the mathematical support of the ternary system to be the most cost-effective system, a significant historical background makes the study of the MVL system (especially the ternary system) more intriguing.

1.4   Historical Background

The historical significance of the MVL system is vast and longstanding. Before the emergence of the semiconductor industry, the electrical switching system used to depend on

Table 1.2 Theoretical variation of the cost and complexity of the circuit and interconnect with respect to radix number, assuming binary system values of 100 [10]

Radix R	Cost C, assuming proportional to $(R \times d)$	Cost C, assuming independent of R	Number of interconnect lines compared to 100 binary lines
2	100	100	100
3	95	63.1	64
4	100	50	50
5	107.9	43.1	44
10	150.5	30.1	31

electromagnetic relays and its' derivatives. In its' simplest forms, these relays were binary. But some multi-position devices were also used in different applications such as telephone systems, railroad traffic control, etc.

With the invention of solid-state devices, the two-state switching devices became widespread. However, some preliminary attempts to establish multi-state devices were still taken. A few examples include Rutz commutating transistor [12], the multiple-frequency oscillator concept of Edson [13], the multi-phase device, namely parametron [14], etc. A ternary computer was built in 1958 by Moscow State University, Russia, which was named SETUN [15]. During that period, several other projects related to ternary logic were undertaken. A complete ternary machine named TERNAC and a software emulator were built in 1973 [16]. After the invention of BJT (Bipolar Junction Transistor) and later the MOS (Metal Oxide Semiconductor) transistors, these early developments in the field of multi-level devices came to cease [10]. However, as the binary logic system is advancing towards a standstill, a renewed interest is given to the multiple-state devices for logic and memory applications.

1.5   Scopes of MVL Technology

Though most of the devices in the real world are in binary nature, the concept of MVL is already introduced in several fields, such as MVL memory, quantum computing, high-speed signaling, etc. The research in the field of MVL creates a path for an unforeseen opportunity in different sectors. This section discusses the scopes of MVL in different real-life applications and systems.

1.5.1 Arithmetic Circuit Design

The implementation of MVL in arithmetic circuit design can be generalized into two major fields: current-mode and voltage-mode. In current mode-based multi-valued logic circuits, different levels are represented by different current values. Here, an analog current summing node is used to calculate the summation and difference of input currents using Kirchhoff's current law. No active or passive elements are needed for these circuits, making the operation very simple [17]. The major disadvantage of current mode MVL circuits is that this type consumes higher power due to the constant flow of currents.

On the other hand, in voltage mode MVL circuits, different logic levels are obtained by different voltage levels, maintained by some non-traditional techniques. Compared to current mode circuits, voltage mode circuits are very power-efficient because very low current flows through the circuits during the stable states. The primary source of power consumed here is during the switching stage.

A general block diagram of an MVL arithmetic circuit design is given in Fig. 1.2. Here, a decoder converts the multi-valued input signal into multiwire binary code. Then, the required logic functions are obtained using pseudo binary logic gates in the second stage. Finally, the binary results are encoded into multi-valued outputs using an encoder at the last step. In this approach, the central part of the circuit uses binary logic gates. In this part, the usual requirements of binary logic gates are significantly reduced, and the interconnects used are short and represent a small load. Moreover, these short connections possess low capacitance values and well-defined fan-out. As a result, the cost of the binary section of the circuit is largely decreased [18]. Another approach of MVL circuits is to use multi-valued logic circuits throughout the process. A lot of research has been conducted on both approaches.

The earliest efforts to implement MVL arithmetic circuits are based on MOSFET technology. Later many new technologies were explored to implement MVL circuits. Some of the examples are Resonant Tunneling Diode (RTD), Single Electron Transistor (SET), Fin Field Effect Transistor (FinFET), Fully Depleted Silicon on Insulator (FDSOI), Quantum Dot Gate Field Effect Transistor (QDGFET), Carbon Nano Tube Field Effect Transistor (CNTFET), Graphene Nano Ribbon Field Effect Transistor (GNRFET), Memristor, etc. An overview of these technologies is provided later in this book. In different literature,

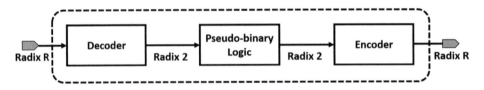

Fig. 1.2 A generalized multi-valued (radix R) functional block [18]

different basic MVL logic gates and arithmetic circuits like the adder, multiplexer, multiplier, and decoder designs are proposed utilizing the above-mentioned technologies. A review of MVL arithmetic circuits using different technologies is available in [19].

1.5.2 Memory Circuit Design

Till date, different MVL memory circuits including SRAM, DRAM, flash memory etc. have been designed using the prevailing technologies like MOSFET and using different emerging technologies. Few of them has also been applied in real-world. The significant technologies that are tried for implementing MVL circuit is discussed in Chap. 3. Also, the various branches of memory circuit which have been tried for MVL representation has been explained in Chaps. 4–7. Among them, Chap. 4 contains the basic design principle of various MVL sequential circuits. Chapter 5 discusses the design principle of both static and dynamic type of MVL Random-Access Memory (RAM). Chapter 6 discusses the design principle of MVL flash memory. Chapter 7 is dedicated for the discussion of Ternary Content Addressable Memory (TCAM). A detailed discussion of all the above-mentioned memory circuits using different technology and different design approaches have been included in the respective chapters of the book.

1.5.3 Quantum Computing

In modern technology, the power dissipation and the overheating issue resulting from the excessive power dissipation causes problems for both the manufacturer and the user. Most of the recent devices work on irreversible logic. As per Landauer's principle [20], irreversible logic computations generate heat energy equal to $kT \times \ln2$ Joules for every bit of lost information (here, k = Boltzmann's constant, T = absolute temperature). Theoretically, reversible logic computation consumes zero power dissipation. Quantum computing ensures that any computation can be implemented in such a way that it is reversible both logically and thermodynamically [21].

The method of using quantum mechanical phenomenon such as interference, superposition, and entanglement to perform computation is called quantum computing [22]. Computers that use quantum computing to solve problems are called quantum computers. In quantum computing, the term "qubit" or quantum bit represents a basic unit of quantum computation. Like a binary bit, a qubit can be represented by two different values (*0* and *1*), but unlike a binary bit, the qubit's state can be a coherent superposition of its' two states. Quantum mechanics can be used to design M-valued quantum gates that can hold multiple-valued information [23]. As multiple-valued quantum computing can characterize an n-dimensional quantum system (based on the states $|0>, |1>, \ldots |n-1>$), it is receiving significant attention in the areas of quantum information theory and quantum cryptography. In a multi-valued quantum system, the unit of information is called "qudit" [21]. The three-valued quantum system is the simplest of the multiple-valued systems, which works with "qutrit."

The Grover search [24] and the factorization algorithm by Shor [25] are famous examples of quantum algorithms that can solve problems in time complexities which is very difficult to be achieved using conventional computing [26]. Furthermore, an IBM quantum computing platform called "IBM Quantum Experience" is available on the cloud, allowing users to work with qubit [27]. Also, recently Google claimed quantum supremacy over other powerful supercomputers by solving a problem using their supercomputer "Sycamore," which is virtually impossible for typical computing systems [28].

1.5.4 High-Speed Signaling

The demand for ultra-high-speed electrical interconnects is increasing day-by-day for intra/inter-chip connections, backplanes, and entire IT infrastructure. Nevertheless, in the case of such high-speed serial links, channel distortions due to ISI (Inter Symbol Interference) resulted from limited channel bandwidth, limiting the I/O bandwidth considerably. As a result, the performance of the total system is hampered. Furthermore, though the advanced MOSFET I/O circuit can utilize the technology developments in terms of switching frequency, the channel bandwidth used for chip-to-chip communication has not been scaled accordingly [29].

One way to increase efficiency in such cases is to communicate using symbols that can receive/transmit more than one bit of data. This principle has been applied for a long time in wireless and wireline communications that employ large complex-valued signal patterns such as Quadrature Amplitude Modulation (QAM) and Quadrature Phase Shift Keying (QPSK) [30]. Lately, this technique has been used to significantly enhance the chip-to-chip and optical communication speed using Pulse Amplitude Modulation (PAM). Recent literature has proposed designs that changed from conventional two-level and non-return-to-zero (NRZ) signaling to 4-level (PAM-4) to obtain throughputs of ≥ 100 Gb/s [18].

1.5.5 Cloud Based Computing Platform

Another highly promising application for MVL is high data processing technology in cloud computing. Cloud computing depends on distributed or shared resources accessible by everyone on the internet. The key objective of the cloud computing platform is to ensure the highest utilization of various distributed resources to ensure maximum throughput and minimum misuse of the resources. The efficiency of cloud computing depends on different performance metrics, each of which can be impacted by many independent

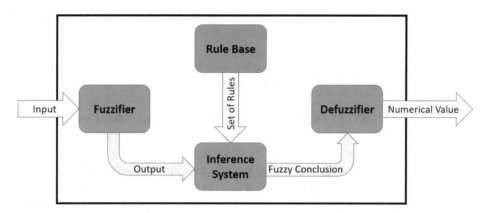

Fig. 1.3 Generalized structure of fuzzy logic control system [31]

parameters. Therefore, a new model is necessary to evaluate the effect of the underlying parameters and integrate those in a unified model theory. Many existing models can be used for this purpose. Fuzzy logic is the most practical solution due to its depth, applicability, and lower complexity. The response of Fuzzy logic to uncertainty and logical reasoning is excellent. Fuzzy logic is used extensively in cloud computing [31].

Fuzzy logic is a form of Multiple-Valued Logic in which the "true" value can be any number between 0 and 1. A fuzzy logic-based computing system works on "degrees of truth" instead of typical Boolean true (1) and false (0) [32]. The concept of Fuzzy logic was instigated by the human decision-making nature, which is benefited by the quality of reasoning with fuzzy or approximate data [33]. This Fuzzy logic theory, which Lotfi Zadeh introduced in 1965, deals with a range of intermediate values between 0 and 1 [34]. It can be used for information related to computational perception and cognition that is uncertain, imprecise, unclear, or without well-defined boundaries [35]. Fuzzy logic is widely used in image processing, data mining, networking, etc. A generalized fuzzy logic control system is shown in Fig. 1.3, consisting of the rule base, fuzzifier, inference system, and defuzzifier [31].

1.5.6   Other Applications of MVL System

Another usefulness of MVL, which has been used for a long time, is the fault detection of binary circuits. One general concept is to use a third state, mainly the intermediate one, to signal a faulty binary circuit. Another way is to use the outer two values for a quaternary (4-valued) system and the innermost value of a quinary (5-valued) system to signal a faulty circuit [36]. Besides the uses mentioned above of MVL techniques, few other uses of MVL have been reported, such as high-throughput and energy-efficient error-control coding techniques [30], signal processing, support chips [36], etc.

1.6   Challenges of MVL System

Multi-Valued Logic (MVL) technologies offer many promising benefits and new applications that cannot be achieved by the conventional binary logic technologies. However, there are many challenges that need to be addressed before MVL devices and circuits can be integrated into any applications. Among the challenges of MVL technologies, the most critical one is the difficulty to maintain three or more logic states defined by distinct voltage levels while the supply voltage (V_{DD}) is going down with technology scaling. The required separations or intermediate voltage gaps between the distinct voltage levels to maintain multiple logic states are affected by noise margins and robustness of the MVL circuit realization techniques against noise disturbances. The commodity MOSFET technology will not be a reliable solution for the MVL devices and circuits. Therefore, new or alternative technologies must be explored.

1.7   Future Direction

Many arithmetic, memory, and sequential circuits have been proposed by researchers using the MVL technologies. This book provides an overview of some of these designs. The direct use of MVL arithmetic circuits in modern industry is yet to be established. But the research in this field creates a path for an unforeseen opportunity. As most of the devices in the real-world are in binary nature, it might take some time to be accustomed to the MVL arithmetic system. In the network router [37] for IPv4 and IPv6, Ternary Content-Addressable Memories (TCAMs) have been used for decades [38]. Generally, the TCAMs are designed using conventional Static Random-Access Memory (SRAM) or Dynamic Random-Access Memory (DRAM) with additional circuitry. Lately, large flash memories having multiple bits per cell have been realized [39]. The notion of communicating data using symbols that carry more than one bit of information has been used for a long time in wireless and in some wireline communication systems, which use large complex-valued signal constellations like Quadrature Amplitude Modulation (QAM) and Quadrature Phase Shift Keying (QPSK). Currently, this technique is being used to increase the speed of chip-to-chip and optical communications by using Pulse Amplitude Modulation (PAM) [39]. In the Ethernet protocol, M-ary PAM is already in use [40]. Another successful commercialization of MVL is the StrataFlash from Intel [41]. In this technology, 2 bits of information can be held in a memory cell using 4 different voltage levels. A NAND-based flash memory was also launched by Samsung that can hold three bits of data in a single cell, which was implemented in the product 840 EVO Series Solid-State-Drive

[42]. A quad-level Dynamic Random-Access Memory (DRAM) was developed by NEC in 1997, which has a capacity of 4 Gb [43, 44]. A quad-level 64 Mb NOR flash memory was developed by STMicroelectronics in 2000 [45]. To address the scaling limitations of planar Multi-Level-Cells (MLC), 3D flash has been proposed by some companies where stacks are stacked vertically. Triple-level memory cells were introduced by Toshiba in 2009 [46]. Samsung also released a type of triple level cells which can store three bits of information per cell with eight different voltage levels [47]. A more detailed discussion on MLC is available in Chap. 6.

1.8   Conclusion

This chapter introduces the definition, evolution, and historical background of Multiple-Valued Logic (MVL). A brief overview of the potential research and implementation scopes of MVL in both present and future systems is provided. For a long time, MVL has been tried by different researchers both in direct and indirect form. In a few fields, the use of MVL is well understood. And in other areas, MVL has only been in the exploration or experimental stages. More time and resources need to be utilized to understand different aspects of MVL and address related challenges. MVL holds the promise of revolutionizing computing and communication technologies by breaking the barriers of binary logic systems.

References

1. S. Tibken, "CES 2019: Moore's Law is dead, says Nvidia's CEO," 2019. [Online]. Available: https://www.cnet.com/news/moores-law-is-dead-nvidias-ceo-jensen-huang-says-at-ces-2019/ [Accessed: 24-Sep-2020].
2. Simonite, T. Intel Puts the Brakes on Moore's Law, MIT Tech. Review, 23 March 2016. https://www.technologyreview.com/s/601102/intel-puts-the-brakes-on-moores-law/ [Accessed: 24-Sep-20 20].
3. https://www.tf.uni-kiel.de/matwis/amat/semitech_en/kap_5/backbone/r5_3_1.html.
4. J. Łukasiewicz, "O Logice trójwartościowej," Ruch Filozoficzny, vol. 5, pp. 170–171, 1920.
5. Post EL. Introduction to a general theory of elementary propositions. American journal of mathematics. 1921 Jul 1;43(3):163–85.
6. Smith, Kenneth C. "The prospects for multi-valued logic: A technology and applications view." IEEE Transactions on Computers 9 (1981): 619–634.
7. Gottwald S, Gottwald PS. A treatise on many-valued logics. Baldock: research studies press; 2001.
8. Wikipedia, the free encyclopedia. "Many-valued logic". [Online]. Available: https://en.wikipedia.org/wiki/Many-valued_logic/ [Accessed: 24-Sep-2020].
9. Donald E Knuth. The art of computer programming, vol 1: Fundamental. Algorithms. Reading, MA: Addison-Wesley, 1968.

10. Stanley L. Hurst. Multiple-valued logic? its status and its future. IEEE Transactions on Computers, (12):1160–1179, 1984.
11. Hallworth, R. P., and F. G. Heath. "Semiconductor circuits for ternary logic." Proceedings of the IEE-Part C: Monographs 109.15 (1962): 219–225.
12. Rutz RF. Two-collector transistor for binary full addition. IBM Journal of Research and Development. 1957 Jul;1(3):212-22.
13. Edson WA. Frequency memory in multi-mode oscillators. IRE Transactions on Circuit Theory. 1955 Mar;2(1):58-66.
14. Komolov VP, Roshal AS. Logic circuits based on ternary parametrons. Soviet Radiophysics. 1965 Jan 1;8(1):129–33.
15. NPMAS BRUSENTSOV. The setun- a small automatic digital computer. 1962.
16. Brian Hayes. Third base. American scientist, 89(6):490–494, 2001.
17. Current KW. Current-mode CMOS multiple-valued logic circuits. IEEE Journal of Solid-State Circuits. 1994 Feb;29(2):95–107.
18. Smith KC. The prospects for multi-valued logic: A technology and applications view. IEEE Transactions on Computers. 1981 Sep 1(9):619–34.
19. Sandhie ZT, Patel JA, Ahmed FU, Chowdhury MH. Investigation of Multiple-valued Logic Technologies for Beyond-binary Era. ACM Computing Surveys (CSUR). 2021 Jan 21;54(1):1–30.
20. Landauer R. Irreversibility and heat generation in the computing process. IBM journal of research and development. 1961 Jul;5(3):183–91.
21. Fan, Fu-You, Guowu Yang, Qian-Qi Le, and Qing-Bin Luo. "A survey of the research on multi-valued quantum circuits." In 2012 International Conference on Wavelet Active Media Technology and Information Processing (ICWAMTIP), pp. 338–341. IEEE, 2012.
22. National Academies of Sciences, Engineering, and Medicine. Quantum computing: progress and prospects. National Academies Press, 2019.
23. Etiemble D. Evolution of Technologies and Multi-valued Circuits. arXiv preprint arXiv:1907.01451. 2019 Jul 1.
24. Shor PW. Algorithms for quantum computation: discrete logarithms and factoring. In Proceedings 35th annual symposium on foundations of computer science 1994 Nov 20 (pp. 124–134). IEEE.
25. Gaudet V. A survey and tutorial on contemporary aspects of multiple-valued logic and its application to microelectronic circuits. IEEE Journal on Emerging and Selected Topics in Circuits and Systems. 2016 Feb 25;6(1):5–12.
26. "IBM Makes Quantum Computing Available on IBM Cloud to Accelerate Innovation", IBM News room, 2016-05-04. https://www03.ibm.com/press/us/en/pressrelease/49661.wss.
27. Tim Childers, "Google's Quantum Computer Just Aced an 'Impossible' Test", Live Science, October, 2019, https://www.livescience.com/google-hits-quantum-supremacy.html/.
28. Grover LK. A fast quantum mechanical algorithm for database search. In Proceedings of the twenty-eighth annual ACM symposium on Theory of computing 1996 Jul 1 (pp. 212–219).
29. Sato N, Chigira T, Toyoda K, Iijima Y, Yuminaka Y. Multi-valued signal generation and measurement for PAM-4 serial-link test. In 2018 IEEE 48th International Symposium on Multiple-Valued Logic (ISMVL) 2018 May 16 (pp. 210–214). IEEE.
30. K. Gopalakrishnan et al., "A 40/50/100 Gb/s PAM-4 Ethernet transceiver in 28 nm CMOS," in IEEE Int. Solid-State Circuits Conf. Dig. Tech. Papers, Feb. 2016, pp. 62–63.
31. Etiemble, D. (2021). Common Fallacies about Multivalued Circuits. *Asian Journal of Research in Computer Science, 12*(4), 67–83. https://doi.org/https://doi.org/10.9734/ajrcos/2021/v12i430295.

32. Chahal RK, Singh S. Trust Calculation Using Fuzzy Logic in Cloud Computing. In Fuzzy Systems: Concepts, Methodologies, Tools, and Applications 2017 (pp. 1314–1366). IGI Global.

33. Zadeh LA. Fuzzy sets as a basis for a theory of possibility. Fuzzy Sets and Systems. 1978 Jan 1;1(1):3-28.

34. Solo AM. The Interdisciplinary Fields of Political Engineering, Public Policy Engineering, Computational Politics, and Computational Public Policy. In Handbook of Research on Politics in the Computer Age 2020 (pp. 1–16). IGI Global.

35. Hayat B, Kim KH, Kim KI. A study on fuzzy logic based cloud computing. Cluster Computing. 2018 Mar;21(1):589-603.

36. Simanjuntak BH, Prasetyo SY, Hartomo KD, Purnomo HD. Application of Fuzzy Logic for Mapping the Agro-Ecological Zones. In Fuzzy Systems: Concepts, Methodologies, Tools, and Applications 2017 (pp. 782–806). IGI Global.

37. Mirzaee, Reza Faghih, and Niloofar Farahani. "Design of a Ternary Edge-Triggered D Flip-Flap-Flop for Multiple-Valued Sequential Logic." arXiv preprint arXiv:1609.03897 (2016).

38. Smith, Kenneth C. "The prospects for multivalued logic: A technology and applications view." IEEE Transactions on Computers 9 (1981): 619–634.

39. Dhande, A. P., and V. T. Ingole. "Design and implementation of 2 bit ternary ALU slice." In Proc. Int. Conf. IEEE-Sci. Electron. Technol. Inf. Telecommun, pp. 17–21. 2005.

40. Atwood G, Fazio A, Mills D, Reaves B. Intel StrataFlash memory technology overview. Intel Technology Journal. 1997;4:1–8.

41. Samsung SSD 840 Series - 3BIT/MLC NAND Flash. [Online]. Available: http://www.samsung.com/global/business/semiconductor/minisite/SSD/uk/html/about/MlcNandFlash.html.

42. Multi-level cell on Wikipedia (https://en.wikipedia.org/wiki/Multi-level_cell).

43. NEC experiments in multi-level cells for DRAMs, by Will Wade on 10.01.1998. Retrieved on 22nd June, 2021 (https://www.eetimes.com/nec-experiments-in-multi-level-cells-for-drams/).

44. STOL (Semiconductor Technology Online), Retrieved on 22nd June, 2021 (http://maltiel-consulting.com/Semiconductor_technology_memory.html).

45. Toshiba Makes Major Advances in NAND Flash Memory with 3-bit-per-cell 32nm generation and with 4-bit-per-cell 43nm technology, 11 Feb, 2009, Retrieved on 22nd June, 2021 (https://www.global.toshiba/ww/news/corporate/2009/02/pr1102.html).

46. Samsung Electronics, Retrieved on 22nd June, 2021 (https://www.samsung.com/us/aboutsamsung/company/history/).

47. R. Drechsler and R. Wille, "Reversible circuits: Recent accomplishments and future challenges for an emerging technology," in Int. Symp. VLSI Design Test, 2012.

Mathematical Representation of Multi Valued Logic

2.1 Definition and Signal Representation

The binary or Boolean numeric system is comprised of two discrete logic levels which is achieved by the ON–OFF switching property of transistors. The two logic levels of a binary system can be expressed as "True (1)" and "False (0)". On the contrary, the multi-valued logic system is a discrete R-valued system, where R > 2. An MVL system can be consisted of set of a certain number of distinct logic levels which are illustrated by signal variables like voltage, current or charge. The logic levels of an MVL system can be discretized using two categories: unbalanced and balanced. In an unbalanced MVL system, the logic levels extend in one particular direction, for example, 0, 1, 2, … (R − 2), (R − 1). And in a balanced MVL system, the logic levels extend in both directions. It can be demonstrated as: (−P), (1 − P) … −1, 0, 1…, (P − 1), P, here R = 2P + 1 [1].

As per the definition mentioned above, a ternary logic system has three different values to signify false, true, and undefined/unknown states [2]. Tables 2.1 and 2.2 represent a voltage-controlled unbalanced (0, 1, 2) and balanced (−1, 0, 1) ternary logic system, respectively. Most recent work focuses on unbalanced logic system as generating two positive voltage levels (V_{DD} and ½V_{DD}) is more realizable than generating one negative (V_{DD}) and one negative (−V_{DD}) voltage level.

2.2 Basic Algebraic Operators for MVL

The General Ternary Inverter (GTI) circuit for an unbalanced ternary logic system can be of three types depending on the output: Negative Ternary Inverter (NTI), Positive Ternary Inverter (PTI) and Standard Ternary Inverter (STI). Equations (2.1)–(2.3) demonstrate the three different kinds of ternary inverters, where x represents the input and y_0, y_1 and y_2

Z. T. Sandhie et al., *Beyond Binary Memory Circuits*, Synthesis Lectures on Digital Circuits & Systems, https://doi.org/10.1007/978-3-031-16195-7_2

Table 2.1 Logic levels representing an unbalanced ternary (3-valued) logic system

Logic level	Logic symbol
0	0
$1/2 \times V_{DD}$	1
V_{DD}	2

Table 2.2 Logic levels representing a balanced ternary (3-valued) logic system

Logic level	Logic symbol
$-V_{DD}$	-1
0	0
V_{DD}	1

are the outputs that represent an NTI, a PTI and an STI respectively [3]. Table 2.3 shows the truth table for the three types of basic inverters.

$$y_0 = C_0(x) = \begin{cases} 2, & x = 0 \\ 0, & x \neq 0 \end{cases} \tag{2.1}$$

$$y_1 = C_1(x) = \begin{cases} 2, & x \neq 2 \\ 0, & x = 2 \end{cases} \tag{2.2}$$

$$y_2 = C_2(x) = \bar{x} = 2 - x \tag{2.3}$$

Some basic MVL operators like Min, Max, Modulo-Sum, Modulo-Difference are shown in Table 2.4.

Table 2.3 Truth table representing three types of ternary inverter (unbalanced)

Input	STI	PTI	NTI
0	2	2	2
1	1	2	0
2	0	0	0

Table 2.4 Basic algebraic operators for MVL [4]

Operator	Notation	Definition
Min	$x \cdot y$	If $x < y$, then x Otherwise y
Max	$x + y$	If $x > y$, then x Otherwise y
Modulo-sum	$x \oplus y$	$(x + y) \mod_R$
Modulo-difference	$x \ominus y$	$(x - y) \mod_R$

Table 2.5 Basic algebraic operators for R = 2, 3 and 4 (considering unbalanced systems) [4, 5]

Min

R=2

·	0	1
0	0	0
1	0	1

R=3

·	0	1	2
0	0	0	0
1	0	1	1
2	0	1	2

R=4

·	0	1	2	3
0	0	0	0	0
1	0	1	1	1
2	0	1	2	2
3	0	1	2	3

Max

R=2

+	0	1
0	0	1
1	1	1

R=3

+	0	1	2
0	0	1	2
1	1	1	2
2	2	2	2

R=4

+	0	1	2	3
0	0	1	2	3
1	1	1	2	3
2	2	2	2	3
3	3	3	3	3

Modulo-sum

R=2

⊕	0	1
0	0	1
1	1	0

R=3

⊕	0	1	2
0	0	1	2
1	1	2	0
2	2	0	1

R=4

⊕	0	1	2	3
0	0	1	2	3
1	1	2	3	0
2	2	3	0	1
3	3	0	1	2

Modulo-difference

R=2

⊖	0	1
0	0	1
1	1	0

R=3

⊖	0	1	2
0	0	2	1
1	1	0	2
2	2	1	0

R=4

⊖	0	1	2	3
0	0	3	2	1
1	1	0	3	2
2	2	1	0	3
3	3	2	1	0

Here, the OR and AND operations are referred by + and · respectively. The details of the operations shown in Table 2.4 is given in Table 2.5.

The "Min", "Max" and "Modulo-Sum" operators described in Tables 2.4 and 2.5 correspond to basic logic operations AND, OR and Ex-OR respectively [4]. From Table 2.5, the truth tables for some unbalanced basic ternary logic gates can be derived as shown in Table 2.6, which demonstrates the truth tables for AND, OR, XOR and Half-Adder. Similarly, the truth tables for the same but balanced basic ternary logic gates can be derived as shown in Table 2.7.

For the design of ternary memory elements, the most common basic gates used are ternary inverters (Flip-Flop, SRAM), ternary NAND (Flip-Flop) and ternary NOR (Flip-Flop).

Table 2.6 Basic logical operators of an unbalanced ternary logic system

Input		OR	AND	Half Adder	
X_1	X_2			Sum	Carry
0	0	0	0	0	0
0	1	1	0	1	0
0	2	2	0	2	0
1	0	1	0	1	0
1	1	1	1	2	0
1	2	2	1	0	1
2	0	2	0	2	0
2	1	2	1	0	1
2	2	2	2	1	1

Table 2.7 Basic logical operators of a balanced ternary logic system

Input		OR	AND	Half Adder	
X_1	X_2			Sum	Carry
-1	-1	-1	-1	-1	-1
-1	0	0	-1	0	-1
-1	1	1	-1	1	-1
0	-1	0	-1	0	-1
0	0	1	0	1	-1
0	1	1	0	-1	0
1	-1	1	-1	1	-1
1	0	1	0	-1	0
1	1	1	1	0	0

2.3   Synthesis Technique of MVL

Logic synthesis is the process in computer engineering, where an abstract design description with certain specification is translated into a gate-level representation. Generally, it can be separated in two different phases: logic optimization and technology mapping. In logic optimization, simplification of the design is done by removing redundancy and detecting the common expressions to reduce the complexity. As this step does not take into account the types of gate that is going to be used in the final design, this step is termed as technology-independent logic optimization. And the technology mapping phase converts

Table 2.8 Karnaugh (K) map for sum and carry operator of a ternary half-adder

Sum				Carry			
X_1/X_2	0	1	2	X_1/X_2	0	1	2
0		1	2	0			
1		2		1			1
2	2		1	2		1	1

the optimized design into a circuit with the help of logic gates available in given target technology [6].

In case of an MVL system, as the radix number increases, the function minimization methods become more complicated. For ternary logic system, the minimization technique is a little complex than that of binary logic but the simplification process of Karnaugh map (K-map) for binary logic can be expanded to ternary logic with some modifications. In [7], a mathematical model to optimize a ternary function using a K-map is offered. From Table 2.6, the K-map table for the sum/carry function of a ternary half adder can be developed as in Table 2.8 and Eqs. (2.4) and (2.5).

The equations for characterizing the sum and carry from the k-map is written as:

$$Sum = 2 \times (X_1^2 X_2^0 + X_1^1 X_2^1 + X_1^0 X_2^2) + 1 \times \left(X_1^1 X_2^0 + X_1^0 X_2^1 + X_1^2 X_2^2\right) \qquad (2.4)$$

$$Carry = 0 + 1 \times \left(X_1^2 X_2^1 + X_1^1 X_2^2 + X_1^2 X_2^2\right) \qquad (2.5)$$

Along with K-map, few other synthesis methods are also available, for example: Galois Field Polynomials, Arithmetic Polynomials and Linear Cellular Arrays [8]. In the case of Galois Field Polynomials, a random logic function f is considered which have m-values and n-variables and the function is illustrated using m^n separate polynomial expressions or polarities. After that, an optimized representation of the function is acquired. The coefficients needed to represent the polynomials can be obtained using direct and inverse Reed-Muller Transforms over Galois Field [7–9]. Arithmetic Polynomial is like Galois Field Polynomials, where along with different polynomials, different polarities can be acquired with the help of direct and inverse arithmetic transform matrices. In the case of Linear Cellular Arrays, the arbitrary logic function f is divided into a number of subvectors. After that the multi-valued variables are encoded into new binary pseudo-variables and the f is characterized by an arithmetic representation built on those pseudo-variables [7, 8]. One of the literatures proposes a decomposition-based mapping method where the synthesis is done by mapping an input matrix to an output matrix [10]. Another approach for calculating and analyzing MVL has been proposed in [8], which can be applied and adapted to any multi-valued function. In this approach, the process is separated into three phases: domain selection, linear regression, and pattern matching for deriving selection criteria.

In addition to the above-mentioned techniques, different other synthesis techniques and tools are present [11, 12]. MVSIS (Multi-Valued Sequential Interactive Synthesis) is an open-source program which is a successor of SIS program, both of which are developed by University of California, Berkeley [13]. ABC is another open-source tool of University of California, Berkeley, which is designed for synthesis and verification of logic circuit [14]. BLIF-MV is another language tool by University of California, Berkeley which is an extension of BLIF (Berkeley Logic Interchange Format). It works as an intermediate language between a high-level HDL (Hardware Description Language) and the synthesis and verification system.

2.4   Conclusion

In this chapter, we provided a short description about Multi-Valued Logic (MVL), and few of its basic mathematical operation like inverter, AND, OR, adder etc. Along with it, we discussed about different synthesis methodology of MVL. MVL synthesis is a separate and vast branch of MVL on which lots of research has been going on for decades. An advanced understanding of MVL operators along with a useful and efficient synthesis technique is an imperative step to design more efficient MVL arithmetic and memory circuits.

References

1. K. C. Smith. 1981. The prospects for multivalued logic: A technology and applications view. IEEE Trans. Comput. C-30, 9 (1981), 619–634.
2. Wikipedia. 2018. What is Three Valued Logic. Retrieved from http://wiki.c2.com/?ThreeValuedLogic.
3. S. Lin, Y. Kim, and F. Lombardi. 2011. CNTFET-based design of ternary logic gates and arithmetic circuits. IEEE Trans. Nanotechnol. 10, 2 (2011), 217–225.
4. Miller, D. Michael, and Mitchell A. Thornton. "Multiple valued logic: Concepts and representations." Synthesis lectures on digital circuits and systems 2.1 (2007): 1–127.
5. Hallworth, R. P., and F. G. Heath. "Semiconductor circuits for ternary logic." Proceedings of the IEE-Part C: Monographs 109.15 (1962): 219–225.
6. Joao Miguel Tavares Severino. 2016. Systems synthesis with multi-value logic (MVL) quaternary logic synthesis. Retrieved January 12, 2021 from https://www.semanticscholar.org/paper/Systems-Synthesis-with-Multi-Value-Logic-%28-MVL-%29-Severino/c21f93b97973b742c3c397ef2671936ef5c76b29?p2df.
7. Vlad P. Shmerko, Svetlana N. Yanushkevich, and Sergey Edward Lyshevski. 2018. Computer Arithmetics for Nanoelectronics. CRC Press.
8. Wafi Danesh and Mostafizur Rahman. 2017. A new approach for multi-valued computing using machine learning. In Proceedings of the IEEE International Conference on Rebooting Computing (ICRC'17). IEEE, 1–7.

9. Zeljko Zilic. 1994. Galois Field Circuits and Realization of Multiple-valued Logic Functions. University of Toronto.

10. M. Abd-El-Barr, G. A. Hamid, and M. N. Hasan. 1998. Synthesis of MVL functions using input and output assignments. IEE Proc. Circ. Devices Syst. 145, 3 (1998), 207–212.

11. Robert K. Brayton and Sunil P. Khatri. 1999. Multi-valued logic synthesis. In Proceedings of the 12th International Conference on VLSI Design. IEEE, 196–205.

12. D. Michael Miller, Gerhard W. Dueck, and Dmitri Maslov. 2004. A synthesis method for MVL reversible logic [multiple value logic]. In Proceedings of the 34th International Symposium on Multiple-valued Logic. IEEE, 74–80.

13. Donald Chai, Jie-Hong Jiang, Yunjian Jiang, Yinghua Li, Alan Mishchenko, and Robert Brayton. 2003. MVSIS 2.0 user's manual. Department of Electrical Engineering and Computer Sciences, University of California, Berkeley.

14. Robert Brayton and Alan Mishchenko. 2010. ABC: An academic industrial-strength verification tool. In Proceedings of the International Conference on Computer Aided Verification. Springer, 24–40.

Overview of Different Technologies for Multiple-Valued Memory

3

3.1 Planar MOSFET Technology

3.1.1 Operating Principle

The Metal–Oxide–Semiconductor Field-Effect-Transistor (MOSFET) is a four-terminal device with drain, gate, source, and body terminals. Often, the body terminal is connected to the source terminal which makes it a three-terminal device. The electrical conductivity of the device is controlled by the voltage applied to the metal gate which is insulated from the semiconductor body using a thin oxide layer. Depending on the working principle, MOSFET can be of two types: depletion mode and enhancement mode. In a depletion-mode MOSFET, the device is generally in a ON state (current conducting) at a zero-gate voltage. They can be used as load resistors in circuit design. On the other hand, enhancement mode MOSFET are generally at an OFF state with a zero-gate voltage. They are the most common switching elements in integrated circuits. Depending on the physical structure of the device, both depletion and enhancement mode MOSFETs can be of two types: n-type (n-channel) and p-type (p-channel). Figure 3.1 shows the classification and symbol for different types of MOSFETs.

In p-channel MOSFET (PMOS) the drain and the source terminals are doped with acceptor atoms. As a result, these terminals become positively charged. The substrate is negatively charged with donor atoms. The current flow results due to the flow of positively charged holes through a positively charged (p-type) channel.

In n-channel MOSFET (NMOS), the drain and source terminals are heavily doped with donor atoms. As a result, these terminals become negatively charged. The substrate is positively charged with acceptor atoms. The current flow results due to the flow of negatively charged electrons through a negatively charged (n-type) channel. Figure 3.2 shows the structural diagram of an N-channel MOSFET.

© The Author(s), under exclusive license to Springer Nature Switzerland AG 2022

Z. T. Sandhie et al., *Beyond Binary Memory Circuits*, Synthesis Lectures on Digital Circuits & Systems, https://doi.org/10.1007/978-3-031-16195-7_3

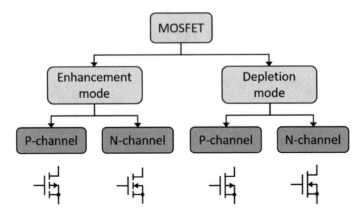

Fig. 3.1 Classification and symbol of MOSFET

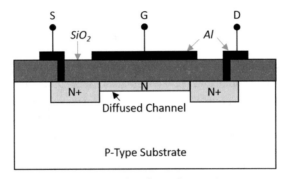

Fig. 3.2 Structure of an N-channel MOSFET

Developing multivalued logic and memory circuits using MOSFET has always been the first choice because the prevailing semiconductor design and fabrication processes are mostly based on MOSFET technology, which makes it simpler, cost effective, and easy to implement in any electronic application. In 1974, Mouftah and Jordan and in later time, other researchers have showed that it is possible to design ternary circuits using MOSFET technology with very little effort and without the help of any new types of transistors [1–5]. The threshold voltage of a MOSFET can be controlled by changing its physical properties during the fabrication process which is one way to produce multiple current or voltage levels required in multivalued devices. Two different approaches have been taken by researchers to implement MVL circuits using MOSFET. The first approach is to use a mixture of enhancement and depletion type MOSFET simultaneously in a circuit [3, 4, 6] and the second approach is to use a combination of enhancement type

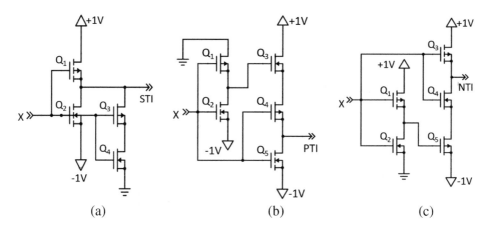

Fig. 3.3 Schematic diagram of **a** STI, **b** PTI, and **c** NTI [3]

of MOSFET [3] and resistors in the same circuit for the realization of multiple voltage levels using voltage divider rule [2, 5].

Figure 3.3 shows the circuit diagrams of a Standard, Negative, and Positive Ternary Inverters (STI, NTI, and PTI) respectively which have been implemented using the first approach [7]. Figure 3.4 shows the design of STI using the second approach. Here, the purpose of the resistors is to obtain an intermediate voltage level using the voltage divider rule [2]. Both these approaches and designs can be extended to implement different ternary logic and arithmetic circuits like AND, OR, Arithmetic Logic Unit (ALU), adder, multiplier, etc. [3, 4, 8, 9].

3.1.2 Analysis

Due to the compatibility with the current processes, MOSFET is the most tested one. Different approaches have been adopted to implement MVL circuit using MOSFETs. Some example implementation techniques are (1) using resistors with MOSFETs, (2) using a combination of both enhancement and depletion mode MOSFETs in the same circuit, (3) threshold voltage variation technique, etc. MVLs can be implemented in current-mode or voltage mode circuits. MOSFETs can be used for both current-mode and voltage-mode MVL circuits.

Although most of the initial works on MVL were mostly based MOSFET technology, recently there are growing interest to use other emerging device technologies for MVL implementation. The main factor that makes MOSFET less appealing for MVL is the Short-Channel Effects (SCE) of MOSFET devices at lower technology nodes. The SCE effects that include carrier velocity saturation, mobility degradation, Drain Induced Barrier Lowering (DIBL), punch-through, and hot carrier effects become more visible when

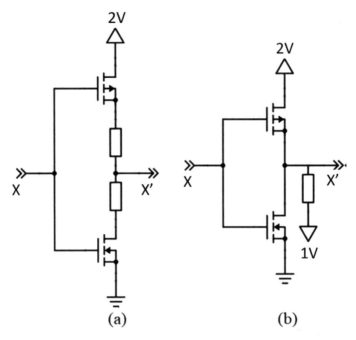

Fig. 3.4 Schematic diagram of an STI design with **a** two resistors **b** one resistor [2]

the length of the transistor channel is comparable to the dimensions of the source and drain regions. SCE effects lead to higher leakage currents, poor ON–OFF characteristics, and uncertainty in threshold voltage levels, which make it very difficult to define and hold multiple voltage levels in MVL circuits. With the shrinking of channel length in MOSFET devices the gate control becomes weaker leading to higher sub-threshold leakage in the channel. The sub-threshold leakage current is one of the key sources of power consumption in current micro and nano electronic devices.

From the design perspective, MOSFET faces a challenge which is to get a stable intermediate voltage at the output node of a ternary/quaternary circuit. For the threshold voltage control method for implementing MVL circuit, it is necessary to have transistors with a wide range of threshold voltage (depending on the supply voltage and number of logic levels used in the design like ternary, quaternary etc.). Though it is possible to control the threshold voltage of a MOSFET certain extent, obtaining a wide range of threshold voltage is not always feasible.

3.2 Silicon on Insulator (SOI) and Fin Field Effect Transistor (FinFET)

To avoid the limitations of the conventional MOSFETs, researchers started focusing on Silicon on Insulator (SOI), Fin Field Effect Transistor (FinFET) and other types of FETs at lower technology nodes that offer lower SCE and better threshold voltage control. In this section, a brief overview of Fully Depleted Silicon-On-Insulator (FDSOI) and FinFET technologies for MVL implementation is presented.

3.2.1 Fully Depleted Silicon-on-Insulator (FDSOI)

The distinctive feature of a SOI MOSFET is the presence of a buried oxide layer, which isolates the body from the substrate. Like bulk MOSFET, SOI MOSFET is a planar device structure and uses similar fabrication process. The main difference is in the silicon wafers that is used for bulk and SOI MOSFET fabrication processes. SOI wafer is comprised of three layers. The first layer of the wafer is a thin layer of silicon on which the transistor is formed. The second layer is an insulating material layer, and the third layer is also made of silicon, which is just physical support layer. The Buried Insulating Oxide (BOX) layer reduces the parasitic junction capacitance, which in turn results in a faster operation. The BOX layer eliminates any leakage path from the gate, as a result the power consumption decreases [10] and the device offers better gate control.

The SOI devices can be implemented in two different types of structures—Partially Depleted (PD) SOI and Fully Depleted (FD) SOI. The FDSOI have a very thin body structures which makes the body completely depleted of any mobile charges. Due to this thin body structure, FDSOI is also called Ultra-Thin-Body Buried Oxide SOI or UTBB-SOI. In case of PDSOI, the body is not as thin as FDSOI making the silicon body partially depleted. The structural diagram of a PDSOI and FDSOI is shown in Fig. 3.5. The advantages of FDSOI are reduced junction capacitance, better threshold voltage variation due to undoped channel, higher electro-static control of the channel, faster switching speed and better performance [11, 12]. Due to these advantages FDSOI are gaining more attention. Unlike bulk MOSFET, an FDSOI device can be controlled by two independent gates leading to better control on channel behavior and threshold voltage. The mechanism to control threshold voltage by using double-gate approach is known as Reverse Body Biasing or RBB [13]. By utilizing the RBB technique it is possible to set different threshold voltage levels and this feature can be exploited to achieve more than two states required to implement MVL circuits.

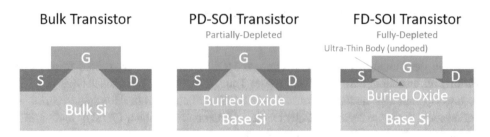

Fig. 3.5 Structure diagram of **a** bulk transistor **b** PD-SOI transistor, and **c** FD-SOI transistor [14]

3.2.2 Fin Field Effect Transistor (FinFET)

The structure of the FinFET allows the fin-shaped channel to be controlled from more than one side by having a wrap-around gate which enables a better control of the channel. A FinFET is a vertical channel device as opposed to the horizontal channel in the planar MOSFET (see Fig. 3.6). The height of the fin in a FinFET is a crucial factor in calculating the channel width (L) as shown in Eq. (3.1).

$$L = 2 \times Fin\ Height + Fin\ Width \tag{3.1}$$

The design and implementation of FinFET based MVL logic circuits are similar to the MVL circuits based on planar MOSFET. However, at lower technology node, FinFET is more dependable and preferable to MOSFET. In FinFET based MVL, the key technique is to control the threshold voltage by controlling the fin properties. There are very few works available in literature where some FinFET based ternary logic and combinational circuits are presented [60–62].

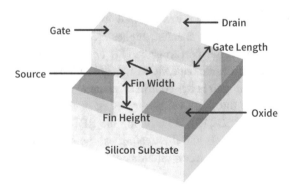

Fig. 3.6 FinFET basic structure [15]

3.2.3 Analysis

FDSOI and FinFET devices are few of the possible solutions to address the limitations of the lower-node planar MOSFET technologies for MVL implementation. Both FDSOI and FinFET offer increased control of the channel and the gate characteristics that result in lower leakage current, higher speed, and better control of the threshold voltage. The fabrication processes for FDSOI and FinFET are also quite matured and already in commercial use.

However, even with many perceived advantages of FDSOI and FinFET over planar MOSFET, the research on FDSOI and FinFET based MVL is still in a very premature stage. Only a handful of basic logic gate designs are proposed by few researchers as of today. Besides, there are some limitations as well. For example, FinFET needs higher number of restricted design rules due to its unique structure. Also, accurate parasitic extraction is more complex in FinFET [16]. On the other hand, FDSOI is susceptible to self-heating as the heat absorbed by the thin body is difficult to dissipate. The increased temperature of the thin body decreases the mobility of the device. Also, it is difficult to fabricate the thin body wafer [17].

Also, in terms of threshold voltage control, though FDSOI and FinFET possess more control than the MOSFET, it is still limited to design a wide range of ternary or quaternary circuits.

3.3 Resonant Tunneling Diode (RTD) Technology

3.3.1 Operating Principle

In a Resonant Tunneling Diode (RTD) device, several semiconducting materials are present in alternative layers. The device consists of two highly doped semiconducting electrodes having little energy gap (e.g., GaAs). These electrodes represent the emitter and collector region. Between these contacts, two barriers comprised of a higher band-gap semiconducting material (e.g., AlGaAs) are present. These barriers contain a quantum well which is also made of lower band-gap material like GaAs. A structure of this kind is called a Double Barrier Quantum Well (DBQW) structure (shown in Fig. 3.7) [18].

In the absence of a forward bias, majority of the electrons and holes generate an accumulation area in the emitter and collector region respectively. When a forward voltage bias is applied, an electric field is created. Due to this electric field, the electrons move from the emitter region to the collector region by tunneling phenomenon inside the quantum well. As a result of the tunneling of electrons across these quasi-bound (open boundary condition at one side and closed boundary condition at another side) energy states, a current takes place. With the increasing voltage, the number of electrons having the same energy in the emitter region increases and more electrons become capable to

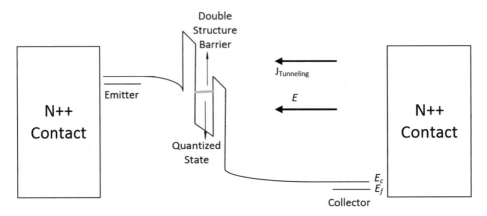

Fig. 3.7 Resonant tunneling diode operation showing the quasi-bound state of the quantum well (Grey) [14]

tunnel through the quantum wall. This ensues an increasing current with the increase of the voltage, which denotes the positive resistance region in Fig. 3.8. Upon reaching a certain point, the energy level of emitter electron and energy level of the quasi-bound state becomes equal. The maximum or peak level of current (I_{ON}) is obtained at that point and the resonant tunneling happens. The occurrence of this phenomenon is dependent on the corresponding doping level and the width of the quantum well. With the increase of the applied voltage, a higher number of electrons gain more energy than the quantum well energy and the currents begin to decrease and reaches the minimum value (I_{VALLEY}). This decrease of current with the increase of voltage results in the Negative-Differential-Resistance (NDR) phenomenon. This is an important characteristic of the RTD device that can be utilized to obtain various voltages for multiple logic levels. Different levels of voltages are obtained in response to the peak and valley currents. After a certain voltage level, the current again increases because of the substantial thermionic emission. In this phase, electrons become capable to tunnel through the non-resonant energy levels as well. This lowest value of current is termed as leakage current. This instance has been simulated in Fig. 3.8, using a simulation tool available in nanoHUB [19].

In literature, different methods for implementing ternary and quaternary circuits using RTD are available. Among these, a very common method is to use a combination of up literal and down literal MOBILE (monostable-to-bistable transition logic elements) to obtain multiple voltage levels at the output [20, 21]. If two RTDs with different I-V characteristics (which result in different threshold voltages) are connected in series combination we get a MOBILE. Depending on the positioning of the two RTDs and the input voltage, two different types of literals (such as: up and down literal) can be designed. The circuit diagrams of an up and a down literal and a ternary inverter design using these are shown in Fig. 3.9.

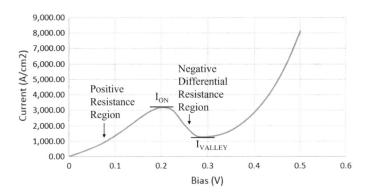

Fig. 3.8 RTD I-V characteristics

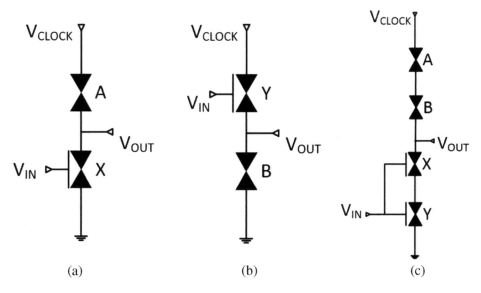

Fig. 3.9 Circuit diagram of **a** down literal MOBILE, **b** up literal MOBILE, and **c** a ternary inverter [20]

In addition to the basic logic gates, different complex logic gates like adder, analog-to-digital converter, pre-decoder etc. based on RTDs have been proposed in [22–24]. Another method to obtain MVL circuits using RTD is to use a combination of RTD along with some other technologies like HEMT (High Electron Mobility Transistors), HBT (Heterojunction Bipolar Transistor), or MODFET (Modulation Doped Field-Effect Transistor) [20, 25].

3.3.2 Analysis

Most of the proposed MVL circuits are implemented using the threshold voltage control property of conventional active devices. These threshold type devices are prone to different limitations like noise margin, and circuit complexity. On the other hand, RTD exhibits some remarkable characteristics like folding I-V curve, and super-fast operation due to the tunneling property. These especial characteristics make RTD an important candidate for MVL applications. Despite these advantages most of the works on RTD based MVL circuits are outdated and there are very few recent works in this field. One of the reasons behind the lack of recent interest in RTD based MVL circuit is the need for a completely different circuit design and implementation technology, for which there is no reliable fabrication process. Due to the absence of matured design and fabrication processes investigating and testing the full potential of RTD based MVL circuit are not possible.

3.4   Single Electron Transistor (SET) Technology

3.4.1 Operating Principle

A Single Electron Transistor (SET) device is composed of three electrodes such as the drain, source, and gate. The drain and the source electrodes are coupled to a Quantum Dot (QD), also known as the island, through a tunnel junction (as shown in Fig. 3.10). The gate electrode is capacitively coupled to the island and can be used to modify the electrical properties of the QD. Figure 3.10 shows the structural diagram of a SET [26].

The operation of a SET is divided into two states—the Blocking State and the Transmitting State, which are defined by the energy levels as shown in Fig. 3.11. In the absence of an external bias, a blocking state arises (Fig. 3.11a). There are no available energy levels in the island, which can be used for tunneling by the source electrons (marked in red). The lower energy levels are pre-occupied. As soon as a positive external bias is applied, island energy levels are lowered (Fig. 3.11b). Hence, the emitter electron (marked in green) can move from initial position "1" through a previously unoccupied energy level in the island (position "2"), and finally reach the drain electrode (position "3"). Here, it scatters inelastically and shifts to the drain electrode Fermi level (position "4"). The gap between the island electrode energy levels is ΔE, which results in a self-capacitance C in the island. The relationship between the capacitance and electrode energy level can be expressed as in (3.2).

$$C = \frac{e^2}{\Delta E} \qquad\qquad (3.2)$$

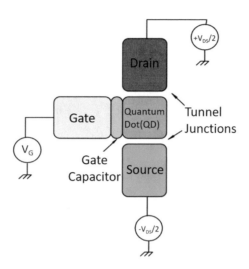

Fig. 3.10 Structural diagram of a single electron transistor [26]

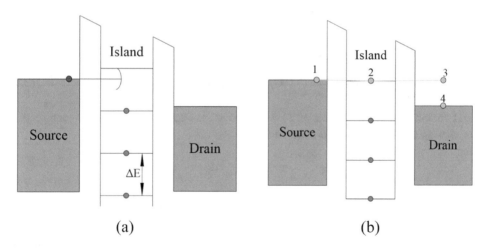

Fig. 3.11   a Blocking state. **b** Transmitting state [26]

SET possesses the capability to include more than one gates which thereby permits the device a property of controllable threshold voltage. A double gate SET device is shown in Fig. 3.12. Here, the gate capacitance of gates 1 and 2 are represented by C_{g1} and C_{g2} respectively, and C_D and C_s are the tunnel junction capacitances of drain and source terminal. One gate (C_{g1}) is used as the input voltage (V_{in}) port and the other gate (C_{g2}) is used for adjusting the threshold voltage of the device. Here, V_{con} is given as the input voltage. The island voltage V_{island} can be represented by (3.3) which controls the current

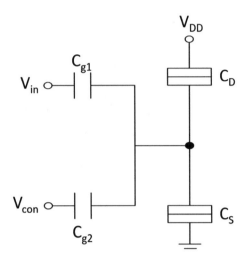

Fig. 3.12 Schematic diagram of dual gate SET [26]

flowing in the two tunnel junctions. Here, n is the number of electrons on the island [27].

$$V_{island} = \frac{1}{C_{g1} + C_{g2} + C_D + C_S}(C_D V_{DS} + C_{g1} V_{in} + C_{g2} V_{con} - ne)$$

$$\Rightarrow V_{island} = \frac{1}{C_\Sigma}(C_D V_{DS} + C_{g1} V_{in} + C_{g2} V_{con} - ne) \qquad (3.3)$$

If the inherent threshold voltage of the island is considered as V_{th}, then the requirement for turning the device on is as shown in (3.4).

$$V_{island} > V_{th}$$

$$\Rightarrow \frac{1}{C_\Sigma}(C_D V_{DS} + C_{g1} V_{in} + C_{g2} V_{con} - ne) > V_{th}$$

$$\Rightarrow V_{in} > \frac{C_\Sigma V_{th} - C_{g2} V_{con} - C_D V_{DS}}{c_{g1}} \qquad (3.4)$$

Therefore, the gate threshold voltage can be expressed by (3.5) [26].

$$V'_{in} > \frac{C_\Sigma V_{th} - C_{g2} V_{con} - C_D V_{DS}}{C_{g1}} \qquad (3.5)$$

By using this concept, basic ternary gates have been proposed in [28] using SET. Two different SET devices with different threshold voltage (V_{th1} and V_{th2}) are used here. The different threshold voltages determine the ternary voltage levels obtained from these gates. Any voltage level which is less than the lower voltage level (for example: V_{th1}) is considered as the logic level "0". Similarly, any voltage level which is higher than V_{th2} (assuming $V_{th2} > V_{th1}$), is considered logic level "2". Finally, an intermediate voltage level is considered to be the logic level "1". A hybrid of SET and MOS can also be used for designing MVL circuits [29].

3.4.2 Analysis

SET offers a way to control threshold voltage of the device by using multiple gates and this is the critical feature to implement MVL circuit. There are some other potential positive features of SET-based MVL circuits. Some of the advantages are compact size, low power consumption, high-speed operation, and ability to integrate with traditional MOS circuits [30]. However, the fabrication process for the SET is still not fully developed. Most of the designs are still in exploratory stage. Fabrication of large quantities of SETs using conventional optical lithography and semiconductor processes and linking the SETs with other circuit components are still very challenging [30]. Also, the current drive capability of SET is very low compared to the MOSFET. SET usually needs to have a low-temperature operation to function properly. For enabling SET to work in normal temperature, single dots having diameters less than 5 nm required to be fabricated and placed between the source and the drain terminals [31]. Another disadvantage of SET is the background charge problem.

3.5   Carbon Based Technologies: CNTFET and GNRFET

3.5.1   Operating Principle

To overcome the scaling issues of silicon based MOSFET devices many emerging materials and device technologies are being explored. Some of the popular emerging materials for the next generation transistors are carbon-based nanomaterials, which have exceptional electrical properties and integration capabilities. Among these, Carbon Nano Tube (CNT) and Graphene Nano Ribbon (GNR) are two forms of graphene, which are widely explored for designing Field Effect Transistors (FETs). There have been many research and experimental works on CNT based FET (CNTFET) and GNR based FET (GNRFET).

Graphene is a two-dimensional allotrope of carbon where the carbon atoms are tightly arranged in a honeycomb lattice. Naturally, graphene is a zero band-gap material. When these graphene sheets are cut into narrow strips having width less than 50 nm, they are called Graphene Nano Ribbon (GNR) [7, 32]. GNR can be categorized into two different categories depending on the edge structure: zigzag and armchair. A zigzag GNR has zero bandgap and always exhibits metallic characteristics. On the contrary, an armchair GNR has a variable bandgap and can show both semiconducting and metallic properties

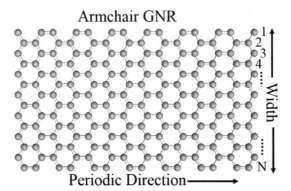

Fig. 3.13 Structure of an armchair GNR with respect to dimer lines [33]

depending on the ribbon width. As a result, armchair GNR is being considered for using in transistor. Dimer number or the number of dimer lines (N) is an important variable while mentioning GNR, which depends on the ribbon width. Figure 3.13 shows the structure of an armchair GNR along with the dimer number. This figure has been drawn using an open-source tool in nanoHUB [33].

The number of dimer lines (N) is a crucial factor in determining the electrical properties of the ribbon. If N can be expressed using the formula $N = 3p$ or $3p + 1$, p being an integer, the GNR acts as a semiconductor. On the other hand, if $N = 3p + 2$, the GNR acts as metallic. In a GNRFET, single or multiple semiconducting armchair GNR layers are used as the channel of the device. Depending on the value of N, the threshold voltage of the device varies. Using this variable threshold voltage characteristics, different MVL logic gates, some complex arithmetic circuits as well as some memory circuits have been proposed [34–37].

When a graphene sheet is rolled up into a hollow, cylindrical structure, a Carbon Nano Tube (CNT) is formed (as shown in Fig. 3.14). Depending on the rolling angle of the graphene sheet, the electrical property of the CNT varies. This rolling angle is known as chirality. The chirality vector is defined by two variables such as: n and m. If the relation between the variables can be defined by the formulae: $n = m$ or $n-m = 3i$ (i being an integer), the CNT exhibits metallic behavior and if $n-m > 3i$, the CNT shows semiconducting nature [38]. Like GNRFET, CNTFET can also be built by using single or multiple CNTs as the channel of the device. By controlling this chirality vector, the threshold voltage of the device can be controlled leading to the ability to achieve multiple logic levels. Some GNRFET and CNTFET based MVL logic and memory gates are proposed in [38–43].

Fig. 3.14   Structure of a chiral CNT [33]

3.5.2   Analysis

In literature, several methods to obtain multiple voltage levels by controlling the threshold voltage of GNRFET and CNTFET have been demonstrated. The design approaches for the GNRFET and CNTFET based MVL circuits are almost interchangeable as both rely on similar techniques to control the threshold voltage. There are few differences among these technologies which are mostly due to the structural and fabrication related issues. These carbon-based transistors are analogous to the conventional MOS transistors, as a result, the MOSFET based designs are convertible to GNRFET and CNTFET based designs. Moreover, they offer higher speed, lower power consumption, and better noise margin. Despite these advantages, the fabrication methodology is still challenging. Line Edge Roughness (LER) is a major issue in fabricating GNR. Lots of research has been going on to perfect the fabrication method to obtain precise and smooth-edged graphene ribbons. Overall, the fabrication of GNRFET is still at an early stage. On the other hand, the fabrication process, and the research on CNTFET are relatively more matured compared to GNRFET. But due to the cylindrical shapes of the CNTs, the design is not transferable to the planar silicon fabrication process.

3.6   Memristor

3.6.1   Operating Principle

Another way of obtaining more than two logic voltage levels is to use resistors as voltage dividers. By selecting appropriate values for these resistors, the desired voltage levels can be obtained. Although, the use of resistors in a circuit is the simplest way to get multiple voltage levels, the solution is not optimal for logic devices because of the large area, bulky size, and high heat dissipation of the conventional resistors. An alternative to the regular resistor is to use memristor, which is considered as the fourth electrical elements. The memristor offers all the usefulness of a resistor like the ability to control or restrict current flow but it requires minimal area and dissipates very low power [44]. Memristor was introduced by Leon Chua in 1971. Before that, resistors, capacitors, and inductors were considered to be the three basic passive electrical components. A resistor defines the relationship between the current and the voltage, a capacitor defines the relationship between the charge and the voltage, and an inductor defines the relationship between the current and the flux. Memristor is the missing passive element which defines the relationship between the flux and the charge and can be defined by (3.6).

$$M = \frac{d\varnothing}{dq} \tag{3.6}$$

Memristor is a non-volatile memory element which does not lose its' data even if the power is cut off. The area consumed by the memristor is very less because it consists of a thin flix (mostly made of TiO_2) sandwiched in between two metal electrodes [45]. The construction of a memristor is shown in Figure 3.15a. And the device symbol of the memristor is shown in Figure 3.15b. The polarity of the device is denoted by the thick black line of the left side of the device. When the current flows into the device, it has a lower value of resistance which can be denoted as R_{on}. On the other hand, when the current flows out of the device, it has a higher value of resistance, which can be denoted by R_{off}.

A current-controlled memristive system can be defined by (3.7), where M denotes the memristance, x defines the state variable and i, q, and t denotes the current, charge and time, respectively [46].

$$v(t) = M(x, q) \times i(t)$$

$$\frac{dx}{dt} = f(x, i(t)) \tag{3.7}$$

Memristor based MVL design is a very new area of research. There is hardly any prior work to compare new ideas and design techniques for memristor based MVL. A

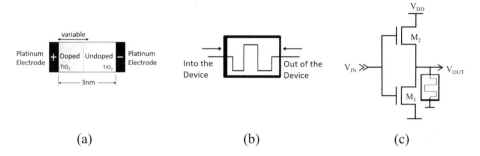

(a)   (b)   (c)

Fig. 3.15 **a** Construction of a memristor [44]; **b** memristive device symbol [47]; **c** ternary inverter using MOSFET and memristor [48]

recently proposed design of a ternary inverter using MOSFET and memristor is shown in Fig. 3.15c [48].

3.6.2 Analysis

Theoretically, any resistive load ternary or other MVL circuits can be re-designed using memristors instead of resistors. The resulting designs will not have the negative features of a resistive design but will offer the positive qualities associated with the memristor. Some recent works used a combination of MOSFET and memristor or CNTFET and memristor to design MVL circuits [48–50]. Although the idea of memristor was presented in 1971, the implementation of a reliable memristive device remained an elusive dream for many years until a potential memristor design was presented as shown in Fig. 3.15 [44]. The use of memristor in MVL circuits is a new concept. More research and experimental work are needed to validate the concept and understand the usefulness and the challenges of using memristor in binary, multivalued, or any logic or memory circuit. The fabrication process and the preferred materials for the memristors are yet to be fully tested and commercialized.

3.7 Magnetic Tunnel Junction (MTJ)

3.7.1 Operating Principle

A Magnetic Tunnel Junction (MTJ) is formed by placing an ultrathin layer of insulator material (e.g., MgO) in between two layers of ferromagnetic material. Due to the thinness of the insulator layer, electrons can tunnel through the insulating barrier if an external voltage is applied between the metal electrodes. Out of these two ferromagnetic layers,

Fig. 3.16 Structure of a
magnetic tunnel junction
(MTJ) [52]

Electrodes → Storage or Free layer
→ Insulator (MgO)
→ Pinned or Fixed layer

one layer is called reference layer or pinned layer and the other layer is called free layer. The magnetization direction of the reference layer is unchangeable and that of the free layer is changeable with the application of the bias voltage. If the magnetization of both layers is parallel, the structure exhibits low resistance R_P. And if the magnetization is anti-parallel, then it has high resistance R_{AP}. The tunneling current is dependent on the relative orientation of these two layers. This phenomenon of spin dependent tunneling is known as the tunneling magnetoresistance (TMR) and can be expressed by (3.8) [51] (Fig. 3.16).

$$TMR = \frac{R_{AP} - R_P}{R_P} \tag{3.8}$$

In MTJ device, various mechanisms are used to obtain switching states. Spin Transfer Torque (STT), Voltage-Controlled Magnetic Anisotropy (VCMA), and Spin–Orbit Torque (SOT) are some examples of mechanisms used to obtain efficient switching states in MTJ devices. A higher TMR state helps easy distinction of different resistance states. A major application of the MTJ devices is MTJ-based memory like Magnetic Random-Access Memory (MRAM), Content Access Memory (CAM), etc. Ternary Content Access Memory (TCAM) is an emerging and important field in the MVL circuit design sector, where the use of a hybrid of MOSFET-MTJ structure is well-established. A detailed discussion of TCAM is included in Chap. 7.

3.7.2 Analysis

The potentials of MTJ devices both in the binary and MVL circuit designs are highly promising and there can be many different types of implementations. A MTJ device has the capability to replace multiple transistors in the circuit [52]. Some of its advantages are non-volatility, high relative magnetoresistance, high scalability, and almost infinite endurance. Due to these positive features MTJs hold an immense promise in the memory sector [53]. It is anticipated that the cell area would not be greatly affected by the MTJ because of its smaller size and the potential ability to stack MTJs over MOS transistors [54]. Other factors that would influence MTJ's switching behavior and magnetic states and the materials to improve the performance of the MTJ devices are currently being investigated. However, the fabrication process of MTJ still requires significant improvement and

Fig. 3.17 Structure of a
neuron-MOS transistor

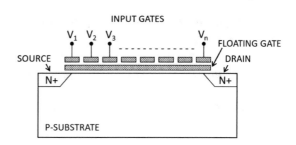

verification before it can be prescribed for actual applications. A very few companies can produce MTJ devices. Some of the challenging and difficult issues of MTJs are preserving high thermal stability and low switching current. The process variations can also result in erroneous reading [52].

3.8   Neuron Metal Oxide Semiconductor (Neuron-MOS)

3.8.1   Operating Principle

The concept of Neuron Metal Oxide Semiconductor (Neuron-MOS) transistor was introduced by Shibata and Ohmi in 1992 [55]. It is also termed as Multiple Input Floating Gate (MIFG) transistor or Floating Gate MOS (FGMOS) [55]. A neuron-MOS transistor is comprised of a floating gate and several input gates. These input gates can interact capacitively with the floating gate. The potential of the floating gate is determined by the control gates. The transistor turns ON when the weighed sum of all the input gates crosses a particular threshold voltage. Due to the enhanced capability of controlling the device current, this type of device can be used for realizing MVL function [57]. The neuron MOS transistors are used in different applications such as variable threshold voltage transistor, single gate Digital-to-Analog Converter, and soft-hardware logic circuit [58]. The structure of a neuron-MOS transistor is shown in Fig. 3.17 [55], where the floating gate and the n-number of control gates are shown. ϕ_F is the floating gate potential and V_1, V_2, ..., V_n are the input voltages of the control gates.

3.8.2   Analysis

Ideally, neuron MOS transistors can be fabricated by any standard MOS transistor process by using polysilicon layer as the floating gate and the first metal layer as the control gates. The only issue arises here is the difference between the dielectric thickness of polysilicon-channel and polysilicon-metal layer. As the capacitance between the floating gate-channel and single weight capacitance between the floating gate-control gate need to be similar, a large difference between the capacitances might result to higher layout areas [59]. One possible solution might be the use of an analog process with two polysilicon layers [56, 59]. The physical and simulation models of the MOS transistor can be used for the Neuron MOS transistor with some small adaptation. The major disadvantages of this technology are lower output resistance, lower effective transconductance, higher area, and uncertainty regarding the amount of charge that might be trapped in the floating gate during the fabrication [56].

3.9   Conclusion

This chapter provides an overview of different technologies that have been studied to implement Multiple Valued Logic (MVL) circuits. Some of the technologies are explored for either arithmetic or memory circuits and some are explored for both types of MVL circuits. To provide a basic understanding of different technologies a brief description of each technology is presented in the relevant sections along with the discussion on the prospects and constraints of these technologies for MVL circuits. Among the technologies that are discussed, some (MOSFET, SOI and FinFET) have already become industry standards for conventional binary logic and memory applications. The rests (such as, CNTFET, Memristor, and Neuron MOS) are still in either conceptual or experimental stages. Majority of the currently available research works focus on multivalued logic circuits and there is a huge gap in multivalued memory related research. In this chapter, most of reliable research on multivalued memory is mentioned.

References

1. H. T. Mouftah and I. B. Jordan. 1974. Implementation of 3-valued logic with cosmos integrated circuits. *Electron. Lett.* 10, 21 (1974), 441–442.
2. X. W. Wu and F. P. Prosser. 1990. CMOS ternary logic circuits. *IEE Proc. G-Circ. Dev. Syst.* 137, 1 (1990), 21–27.
3. Alex Heung and H. T. Mouftah. 1985. Depletion/enhancement CMOS for a lower power family of three-valued logic circuits. *IEEE J. Solid-State Circ.* 20, 2 (1985), 609–616.
4. Prabhakara C. Balla and Andreas Antoniou. 1984. Low power dissipation MOS ternary logic family. *IEEE J. Solid-State Circ.* 19, 5 (1984), 739–749.
5. H. T. Mouftah and K. C. Smith. 1982. Injected voltage low-power CMOS for 3-valued logic. In *IEE Proceedings G-Electronic Circuits and Systems*, Vol. 129. IET, 270–272.
6. E. Kinvi-Boh, M. Aline, Olivier Sentieys, and Edgar D. Olson. 2003. MVL circuit design and characterization at the transistor level using SUS-LOC. In *Proceedings of the 33rd International Symposium on Multiple-Valued Logic*. IEEE, 105–110.
7. Wikipedia. Graphene Nanoribbon. Retrieved January 12, 2021 from https://en.wikipedia.org/wiki/Graphene_nanoribbon.

8. A. P. Dhande and V. T. Ingole. 2005. Design and implementation of 2 bit ternary ALU slice. In *Proceedings of the International IEEE Conference on Electronics Technology and Information Telecommunication.* 17–21.

9. A. Srivastava and K. Venkatapathy. 1996. Design and implementation of a low power ternary full adder. *VLSI Design* 4, 1 (1996), 75–81.

10. Design & Reuse. 2018. "CMOS-SOI-FinFET". Retrieved from https://www.design-reuse.com/articles/41330/cmos-soifinfet-technology-review-paper.html.

11. ST. 2018. FDSOI. Retrieved from https://www.st.com/content/st_com/en/about/innovation---technology/FD-SOI/learn-more-about-fd-soi.html.

12. Wei HW, Ruslan SH. Investigation of FDSOI and PDSOI MOSFET characteristics. In AIP Conference Proceedings 2019 Nov 11 (Vol. 2173, No. 1, p. 020005). AIP Publishing LLC.

13. Sumanta Chaudhuri, Tarik Graba, and Yves Mathieu. 2016. Multivalued routing tracks for FPGAs in 28nm FDSOI technology. Retrieved from arXiv:1609.08681.

14. Olejarz P, Park K, MacNaughton S, Dokmeci MR, Sonkusale S. 0.5 µW Sub-Threshold Operational Transconductance Amplifiers Using 0.15 µm Fully Depleted Silicon-on-Insulator (FDSOI) Process. Journal of Low Power Electronics and Applications. 2012 Jun;2(2):155–67.

15. CircuitBread 2021, "What is a FinFET?", Retrieved from https://www.circuitbread.com/ee-faq/what-isa-finfet.

16. SignOff. 2018. FinFET. Retrieved from http://www.signoffsemi.com/finfet/.

17. Design & Reuse. 2018. CMOS-SOI-FinFET. Retrieved from https://www.design-reuse.com/articles/41330/cmos-soifinfet-technology-review-paper.html.

18. Johnny Ling. 1999. Resonant tunneling diodes: Theory of operation and applications. University of Rochester, Rochester, NY.

19. Hong-Hyun Park, Zhengping Jiang, Arun Goud Akkala, Sebastian Steiger, Michael Povolotskyi, Tillmann Christoph Kubis, Jean Michel D. Sellier, Yaohua Tan, SungGeun Kim, Mathieu Luisier, et al. 2008. Resonant Tunneling Diode Simulation with NEGF. Retrieved on January 12, 2021 from https://nanohub.org/tools/rtdnegf.

20. Takao Waho, Kevin J. Chen, and Masafumi Yamamoto. 1998. Resonant-tunneling diode and HEMT logic circuits with multiple thresholds and multilevel output. *IEEE J. Solid-State Circ.* 33, 2 (1998), 268–274.

21. Mi Lin and Ling-ling Sun. 2012. A novel ternary JK flip-flop using the resonant tunneling diode literal circuit. *J. Zhejiang Univ. Sci. C* 13, 12 (2012), 944–950.

22. H. C. Lin. 1994. Resonant tunneling diodes for multivalued digital applications. In *Proceedings of the 24th International Symposium on Multiple-Valued Logic (ISMVL'94).* IEEE, 188–195.

23. Federico Capasso, Susanta Sen, Fabio Beltram, Leda M. Lunardi, Arvind S. Vengurlekar, P. R. Smith, N. J. Shah, Roger J. Malik, and A. Y. Cho. 1989. Quantum functional devices: Resonant-tunneling transistors, circuits with reduced complexity, and multiple valued logic. *IEEE Trans. Electron Devices* 36, 10 (1989), 2065–2082.

24. Hui Zhang, Tetsuya Uemura, Pinaki Mazumder, and Kyounghoon Yang. 2004. Design of multivalued QMOS predecoder. In *Proceedings of the 4th IEEE Conference on Nanotechnology.* IEEE, 614–617.

25. Pinaki Mazumder, Shriram Kulkarni, Mayukh Bhattacharya, Jian Ping Sun, and George I. Haddad. 1998. Digital circuit applications of resonant tunneling devices. *Proc. IEEE* 86, 4 (1998), 664–686.

26. Wikipedia. 2021. Coulomb Blockade. Retrieved from https://en.wikipedia.org/wiki/Coulomb_blockade.

27. Peter Hadley, Gunther Lientschnig, and Ming-Jiunn Lai. 2003. Single-electron transistors. In *Proceedings of the Conference Series—Institute of Physics*, Vol. 174. Institute of Physics, Philadelphia, PA, 125–132.

28. Wu Gang, Cai Li, and Li Qin. 2009. Ternary logic circuit design based on single electron transistors. *J. Semiconduct.* 30, 2 (2009), 025011.
29. Hiroshi Inokawa, Akira Fujiwara, and Yasuo Takahashi. 2001. A multiple-valued logic with merged single-electron and MOS transistors. In *Proceedings of the International Electron Devices Meeting.* IEEE, 7–2.
30. Monika Gupta. 2016. A study of single electron transistor (SET). *Int. J. Sci. Res* 5, 1 (2016), 474–479.
31. Klüpfel FJ, Burenkov A, Lorenz J. Simulation of silicon-dot-based single-electron memory devices. In 2016 International Conference on Simulation of Semiconductor Processes and Devices (SISPAD) 2016 Sep 6 (pp. 237–240). IEEE.
32. Mohammed, Mahmood Uddin, and Masud H. Chowdhury. "Design of Energy Efficient SRAM Cell Based on Double Gate Schottky-Barrier-Type GNRFET with Minimum Dimer Lines." 2019 IEEE International Symposium on Circuits and Systems (ISCAS). IEEE, 2019.
33. Gyungseon Seol, Youngki Yoon, James K Fodor, Jing Guo, Akira Matsudaira, Diego Kienle, Gengchiau Liang, Gerhard Klimeck, Mark Lundstrom, Ahmed Ibrahim Saeed (2019), "CNT-bands," https://nanohub.org/resources/cntbands-ext. (DOI: https://doi.org/10.21981/QT2F-0B32).
34. Sandhie ZT, Ahmed FU, Chowdhury M. GNRFET based ternary logic–prospects and potential implementation. In 2020 IEEE 11th Latin American Symposium on Circuits & Systems (LASCAS) 2020 Feb 25 (pp. 1–4). IEEE.
35. Sandhie ZT, Ahmed FU, Chowdhury MH. Design of Ternary Logic and Arithmetic Circuits using GNRFET. IEEE Open Journal of Nanotechnology. 2020 Sep 1;1:77–87.
36. Divya Geethakumari Anil, Yu Bai, and Yoonsuk Choi. 2018. Performance evaluation of ternary computation in SRAM design using graphene nanoribbon field effect transistors. In *Proceedings of the IEEE 8th Annual Computing and Communication Workshop and Conference (CCWC'18).* IEEE, 382–388.
37. Zarin Tasnim Sandhie, Farid Uddin Ahmed, and Masud H. Chowdhury. 2020. Design of ternary master-slave D-flip flop using MOS-GNRFET. In *Proceedings of the IEEE 63rd International Midwest Symposium on Circuits and Systems (MWSCAS'20).* IEEE, 554–557.
38. Lin, Sheng, Yong-Bin Kim, and Fabrizio Lombardi. "CNTFET-based design of ternary logic gates and arithmetic circuits." IEEE transactions on nanotechnology 10, no. 2 (2011): 217–225.
39. Jaber RA, Kassem A, El-Hajj AM, El-Nimri LA, Haidar AM. High-performance and energy-efficient CNFET-based designs for ternary logic circuits. IEEE Access. 2019 Jul 11;7:93871–86.
40. Zahoor F, Hussin FA, Khanday FA, Ahmad MR, Mohd Nawi I, Ooi CY, Rokhani FZ. Carbon Nanotube Field Effect Transistor (CNTFET) and Resistive Random-Access Memory (RRAM) Based Ternary Combinational Logic Circuits. Electronics 2021, 10, 79.
41. Jaber RA, Owaidat B, Kassem A, Haidar AM. A Novel Low-Energy CNTFET-Based Ternary Half-Adder Design using Unary Operators. In 2020 International Conference on Innovation and Intelligence for Informatics, Computing and Technologies (3ICT) 2020 Dec 20 (pp. 1–6). IEEE.
42. Jaber RA, El-Hajj AM, Kassem A, Nimri LA, Haidar AM. CNFET-based designs of Ternary Half-Adder using a novel "decoder-less" ternary multiplexer based on unary operators. Micro-electronics Journal. 2020 Feb 1;96:104698.
43. Mohammad Hossein Moaiyeri, Reza Faghih Mirzaee, Akbar Doostaregan, Keivan Navi, and Omid Hashemipour. 2013. A universal method for designing low-power carbon nanotube FET-based multiple-valued logic circuits. *IET Comput. Dig. Techn.* 7, 4 (2013), 167–181.
44. edgefx.in, Retrieved from: https://www.edgefx.in/memristor-know-memristor-technology/#:~:text=The%20following%20are%20the%20benefits,are%20comfortable%20with%CMOS%20interfaces.

45. WatElectronics.com, Retrieved from: https://www.watelectronics.com/what-is-memristor-con struction-its-working/

46. Alharbi AG, Fouda ME, Chowdhury MH. Memristor emulator based on practical current controlled model. In 2015 IEEE 58th International Midwest Symposium on Circuits and Systems (MWSCAS) 2015 Aug 2 (pp. 1–4). IEEE.

47. Johnsen GK. An introduction to the memristor-a valuable circuit element in bioelectricity and bioimpedance. Journal of Electrical Bioimpedance. 2012 Aug 9;3(1):20–8.

48. Muhammad Khalid and Jawar Singh. 2016. Memristor-based unbalanced ternary logic gates. *Analog Integr. Circ. Signal Process.* 87, 3 (2016), 399–406.

49. A. A. El-Slehdar, A. H. Fouad, and A. G. Radwan. 2013. Memristor-based balanced ternary adder. In *Proceedings of the 25th International Conference on Microelectronics (ICM'13)*. IEEE, 1–4.

50. Mahmood Uddin Mohammed, Rakesh Vijjapuram, and Masud H. Chowdhury. 2018. Novel CNTFET and memristor based unbalanced ternary logic gate. In *Proceedings of the IEEE 61st International Midwest Symposium on Circuits and Systems (MWSCAS'18)*. IEEE, 1106–1109.

51. Zhu JG, Park C. Magnetic tunnel junctions. Materials today. 2006 Nov 1;9(11):36–45.

52. Maciel N, Marques E, Naviner L, Zhou Y, Cai H. Magnetic tunnel junction applications. Sensors. 2020 Jan;20(1):121.

53. Peng SZ, Zhang Y, Wang MX, Zhang YG, Zhao W. Magnetic tunnel junctions for spintronics: principles and applications. Wiley Encyclopedia of Electrical and Electronics Engineering. 1999 Dec 27:1–6.

54. Matsunaga S, Miura S, Honjou H, Kinoshita K, Ikeda S, Endoh T, Ohno H, Hanyu T. A 3.14 um 2 4T-2MTJ-cell fully parallel TCAM based on nonvolatile logic-in-memory architecture. In2012 Symposium on VLSI Circuits (VLSIC) 2012 Jun 13 (pp. 44–45). IEEE.

55. Shibata T, Ohmi T. A functional MOS transistor featuring gate-level weighted sum and threshold operations. IEEE Transactions on Electron devices. 1992 Jun;39(6):1444–55.

56. Rodriguez-Villegas, E., 2006. *Low power and low voltage circuit design with the FGMOS transistor* (Vol. 20). IET.

57. Shen J, Tanno K, Ishizuka O, Tang Z. Application of neuron-MOS to current-mode multi-valued logic circuits. In Proceedings. 1998 28th IEEE International Symposium on Multiple-Valued Logic (Cat. No. 98CB36138) 1998 May 29 (pp. 128–133). IEEE.

58. Pethe SH, Narkhede S. Design of Ternary D Flip-Flop Using Neuron MOSFET., IOSR Journal of VLSI and Signal Processing (IOSR-JVSP) Volume 5, Issue 2, Ver. II (Mar. - Apr. 2015), PP 36–39.

59. Weber W, Prange SJ, Thewes P, Wohlrab E, Luck A. On the application of the neuron MOS transistor principle for modern VLSI design. IEEE transactions on Electron Devices. 1996 Oct;43(10):1700–8.

60. Kishor MN, Narkhede SS. A novel finfet based approach for the realization of ternary gates.

61. Kishor MN, Narkhede SS. Design of a ternary finfet sram cell. In 2016 Symposium on Colossal Data Analysis and Networking (CDAN), pages 1–5. IEEE, 2016.

62. Jyoti K, Narkhede S. An approach to ternary logic gates using finfet. In Proceedings of the International Conference on Advances in Information Communication Technology & Computing, pages 1–6, 2016.

MVL Sequential Circuits

4

4.1   Introduction

Digital circuits can be divided into two categories defending on the input–output relationship. These two categories are *Combinational Logic Circuit* and *Sequential Logic Circuit*. In a combinational circuit, the output depends on the current states of the inputs. Here, no feedback path from the output to the input is present. As a result, this type of circuit is also known as nonregenerative circuit. In a sequential circuit, the output is dependent on the current inputs as well as the previous values of the output stored in a memory unit. This is achieved by adding the memory unit in a positive feedback path from the output to the input. The function of the memory unit in the feedback path is to remember past events. Therefore, each sequential logic circuit is made of a combinational logic circuit and a memory unit. Sequential circuit is also known as regenerative circuit. Equations (4.1) and (4.2) and Fig. 4.1 show the input–output relations and functional block diagrams of combinational and sequential logic circuits.

$$\text{Combinational Circuit: } Output = f(Input) \tag{4.1}$$

$$\text{Sequential Circuit: } Output = f(Input, Previous\ input) \tag{4.2}$$

Depending on the operating principle, the sequential logic circuits can be subdivided in to two types: Asynchronous and Synchronous circuits. The output in an asynchronous sequential circuit can change at any time depending on the change in the input. On the other hand, the output in a synchronous sequential circuit can change only at certain time depending on a periodic clock signal.

Z. T. Sandhie et al., *Beyond Binary Memory Circuits*, Synthesis Lectures
on Digital Circuits & Systems, https://doi.org/10.1007/978-3-031-16195-7_4

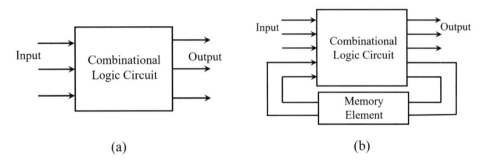

Fig. 4.1 **a** Combinational and **b** sequential logic circuit

Usually, the combinational logic circuit part in the sequential circuit (Fig. 4.1b), is comprised of some basic logic gates like NAND, NOR etc. As a result, a major contributing factor while designing a ternary sequential circuit is the development of basic ternary logic gates. The research on ternary sequential logic circuits started in 1980s. Some of the earlier work on ternary sequential circuits can be found in [1–3].

4.2   Ternary D-Latch

D-Latch is a popular and widely used memory element in integrated circuit designs. Conventional binary D-Latch can store one of the two possible pieces of information or signals at a time and each of these two pieces of information or signals is known as a binary "bit". A ternary memory element can store or retain one of the three possible pieces of information or signals and each of these signals or information is known as a ternary "trit". Therefore, a ternary D-Latch can store one "trit" at a certain time period until the content is replaced/substituted by a new "trit". A basic ternary memory cell can hold three distinct stable states (trits) denoted by 0, 1, and 2. In a latch, the information can be kept in a latch/lock mode or open mode depending on the requirement. A latch is a level-sensitive memory element, where the output is continuously affected by the input signal if the enable signal is turned on. The input signal is referred as "D" in a D-latch. It has two different operating modes - transparent and hold modes.

Table 4.1 represents the truth table of a ternary D-latch with a binary clock/enable signal. When the clock/enable signal value is held at a lower voltage, the device stays in a transparent mode, and the output follows the input signal. When the clock/enable signal is held at higher voltage level, the device remains at a hold mode, and the output holds the previous value. This whole operation can be represented by (4.3).

$$Q = D \times clk + Q_{previous} \times \overline{clk} \qquad (4.3)$$

Table 4.1 Truth table representing a ternary D-latch

En	D	Q	$\overline{Q}$
0	0	0	2
0	1	1	1
0	2	2	0
2	X	Q_{prev}	$\overline{Q}_{prev}$

Like the binary latches, two different approaches can be used to design ternary latches. The two techniques are NAND/NOR based latch design and transmission gate-based latch designs. NAND/NOR based ternary latch requires a larger number of transistors. For a regular binary NAND/NOR gate, we need four (4) transistors. On the other hand, for a ternary NAND/NAR gate we need more than four (4) transistors. In some stable and noise-immune ternary NAND/NOR designs, up to ten (10) transistors are used. A higher number of transistors are required to ensure the correct discretization of multiple levels of the output voltage and to enable the output to hold different logic levels in the presence of larger noise contents. In [4–6], some designs to safely generate the correct outputs in a ternary NAND gate in the presence of larger noise range are demonstrated. Figure 4.2 [6] shows a NAND-based ternary D-latch, which can have three possible outputs: logic "0", logic '1", and logic "2". If two of the inputs of this NAND-based ternary D-latch are logic "2" (V_{DD}), the output should be logic "0". If the V_{DD} is 0.9 V, the circuit can detect the original inputs for a value of ($V_{DD} \pm 30\%$ of V_{DD}). It means that due to noise induced variation if one of the inputs goes down to ~0.6 V the circuit will still be able to produce the correct output (logic "0″).

Figure 4.3 [7] shows a transmission gate-based ternary D-latch. In the transmission-gate based design, the required number of transistors is significantly less than the NAND gate-based design shown in Fig. 4.2. However, the operation of the transmission gate-based design is analog in nature, which means that variations in input will affect the output depending on the gate voltage. The stability of the design depends on the stability of the ternary inverter or buffer used in the circuit.

Fig. 4.2 A NAND gate based ternary D-latch [6]

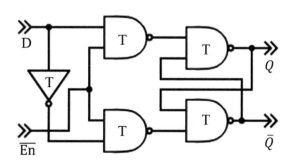

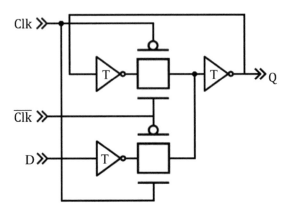

Fig. 4.3 A transmission gate and inverter based ternary D-latch [7]

Another pass-transistor based design of ternary D-latch is presented in [8], which uses a novel ternary inverter/buffer cell. In this design, the ternary D-latch function of (4.3) is implemented by using a pair of transmission gate complementary switches and a ternary inverter/buffer cell with a dual rail structure as shown in Fig. 4.4 [8]. The input signal for the latch is D and $\overline{D}$. When the Clk = logic "2", the inputs are passed to the output and when the Clk = logic "0", the output remains unchanged.

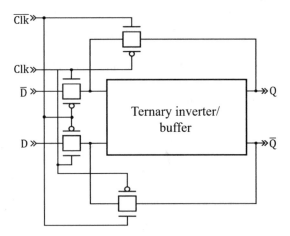

Fig. 4.4 A transmission-gate based ternary D-latch design with a dual rail structure [8]

4.3   Ternary D Flip-Flop

A flip-flop is an edge-triggered memory element unlike a latch, which is a level-sensitive memory element. A flip-flop can either be a positive or negative edge triggered memory circuit. Usually, flip-flops are formed by cascading two opposite (positive and negative) types of latches back-to-back. A ternary flip-flop is also known as a flip-flap-flop due to its' ability to store three distinct logic values [9]. The basic function of a flip-flop (FF) is to store or hold data. Flip-flops are also used to control the operation of digital circuits based on the states stored in one or more FFs [10]. A ternary D flip-flap-flop (DFFF) operates either in transparent or hold mode depending on the triggering edge (positive or negative edge) of the clock signal. A DFFF is a cascade of a positive and a negative ternary latch.

4.3.1   Positive Edge Triggered D Flip-Flap-Flop

A positive edge-triggered D flip-flap-flop (DFFF) goes into the transparent mode at the positive edge of the clock signal. Common techniques for binary flip-flop can be extended for ternary flip-flops. Figure 4.5 shows the structural diagram of a positive edge triggered ternary DFFF, which is made of two NAND based ternary D-latches. Here, the first latch (left) is termed as the master latch and the second latch (right) as the slave latch. When clock signal is low, the master latch is in the transparent mode and the input signal at D is passed to the middle node Q_M. During that time (when clock is low), the slave latch remains at the hold mode and retaining the previous value by its feedback mechanism. During the transition of the clock signal from the low to high level (upon arrival of a positive clock edge) the master latch goes into the hold mode and preserves the value stored at Q_M by feedback. At the same time, the slave latch enters the transparent mode and the value stored at the node Q_M is transmitted to the output node Q. Here, the output Q makes one transition after the arrival of the positive edge of the clock. This cascaded structure of two latches acts a positive edge triggered DFFF. The operation of the ternary DFFF can be represented by the truth table shown in Table 4.2.

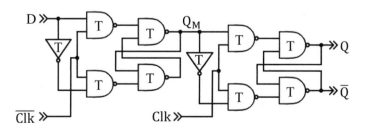

Fig. 4.5   Schematic diagram of a ternary master–slave DFFF using ternary NAND [6, 11–13]

Table 4.2 Truth table representing a ternary D-flip flop

Clock	Mode	D	Q	$\overline{Q}$
$0 \rightarrow 2$	Transparent	0	0	2
		1	1	2
		2	2	0
$2 \rightarrow 0$	Hold	X	Q_{prev}	$\overline{Q}_{prev}$

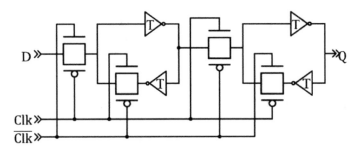

Fig. 4.6 Schematic diagram of a ternary master–slave DFFF using ternary inverter and transmission gate [14]

The ternary DFFF can also be implemented using transmission gates. Figure 4.6 [14] shows the schematic diagram of a ternary master–slave DFFF based on transmission gates.

Another different approach of ternary master–slave flip-flop can be found in [9]. Here, a unique design of ternary latch is first proposed and then using that ternary latch, the master–slave FFF is proposed. Figure 4.7 shows the design concept of the ternary latch and its' symbol. The key difference of the latch design with other prevailing latch design is the output voltage level of the Q and $\overline{Q}$. Usually, these two outputs of a latch are complementary to each other, whereas in the proposed design of [9], these two values are not complementary. The ternary output Q is the average value of the binary output Q_1 and Q_2.

The proposed ternary latch then can be converted to a FFF using a clock signal (Fig. 4.8a). Later, using the proposed level-sensitive FFF, a master–slave FFF is designed (Fig. 4.8c). For the detailed design description, please refer to [8].

As we can see, in all the above-mentioned master–slave FFF designs, a feedback loop is present to hold the values stored on the output nodes of the latches until the rising edge (falling edge in case of a negative flip-flop) of the clock signal arrives. In addition to the ternary DFFF discussed in this section, there are other types of ternary SR flip-flop and JKL flip-flop designs available in literature.

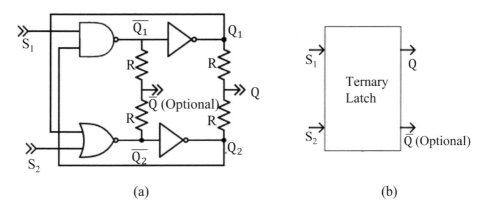

(a)   (b)

Fig. 4.7 **a** Schematic diagram, and **b** symbol of the latch using binary NAND, NOR and inverter [9]

4.4 Analysis

The operation and functional characteristics of ternary sequential circuits are similar to the convention of binary sequential circuits. The proposed designs for ternary D latches, DFFFs, master–slave ternary flip-flops, and other types of ternary flip-flops like JK flip-flop can be implemented using MOSFET, CNTFET, GNRFET and other device technologies. In [2, 9, 11–13, 15] different MOSFET based designs of ternary latches and flip-flops are presented. Some CNTFET and GNRFET based ternary sequential circuits are shown in [6, 16–18]. Some alternative designs based on other emerging device technologies like Single Electron Transistor (SET), Neuron-CMOS, Multiple-Junction Surface Tunnel Transistors, and Resonant Tunneling Diode can be found in [14, 19–22].

To provide an in-depth view of different designs of ternary sequential circuits based on MOSFET, CNTFET, and GNRFET a comparative analysis of power and delay of some proposed ternary edge-triggered flip-flops is presented in Table 4.3. A ternary DFFF can have six possible transitions at the output. Here, the delay comparison is done for all six possible transitions of the output voltage (logic level) in response to the changes of the clock pulse. The simulation results (power-delay information) for the designs presented in literature [7, 12] are taken from literature [9].

The similarities between multi-valued and binary sequential circuits make the understanding and adaptation of the MVL logic more straightforward. From Table 4.3 it is observed that the standby power for storing "1" is exceptionally high. These is because, whenever a ternary logic gate is holding an intermediated value ("1"), both the pull-up and pull-down networks are active. As a result, a path is present between the (V_{DD}) and the ground (GND) lines for the entire duration of the logic level "1". This direct path leads to significant increase of the short-circuit power consumption. This issue will be present in

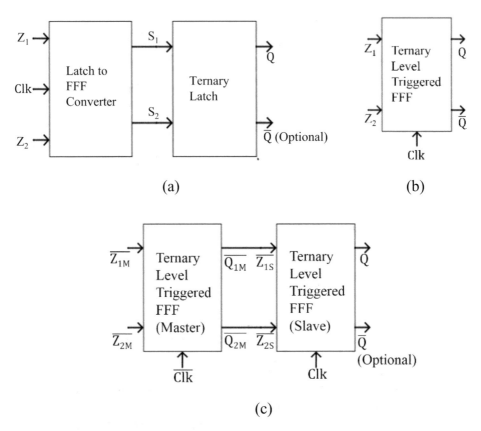

Fig. 4.8 **a** Schematic diagram of ternary level-sensitive flip-flap-flop, **b** symbol of the proposed level-sensitive flip-flap-flop, and **c** ternary edge-sensitive FFF using the master–slave structure [9]

any ternary sequential circuit based on the basic ternary logic gates like inverter, NAND, and NOR.

In some MOSFET based ternary sequential circuit designs, voltage dividers based on basic resistors are used to achieve the intermediate voltage (logic) levels in ternary or multi-valued logic gates. However, the conventional resistors are bulky and consume a lot of area in the layout, and as a result it is not a desirable solution. The potential use of emerging technologies along with the CMOS devices in the MVL circuits will introduce new challenges. The carbon-based devices like CNTFET and GNRFET offer the unique advantage of the ability to smoothly control the threshold voltage to design basic ternary logic gates. However, the commercial and reliable fabrication processes for these novel devices are still at the exploratory stage. The non-planar structure of Carbon Nano Tube (CNT) channel in CNTFET is not compatible to the contemporary MOSFET fabrication process. In case of GNRFET, creating a Graphene Nanoribbon (GNR) without any edge

Table 4.3 Power and delay analysis in different edge-triggered ternary flip-flops

Technology		[9]	[12]	[7]	[6]
		45 nm MOSFET	45 nm MOSFET	32 nm CNTFET	16 nm GNRFET
Delay (ps)	$0 \rightarrow 1$	69.196	90.504	99.416	24.8
	$0 \rightarrow 2$	116.66	326.88	284.92	61
	$1 \rightarrow 0$	61.687	463.01	116.62	111
	$1 \rightarrow 2$	83.393	305.79	232.14	79.9
	$2 \rightarrow 0$	95.044	489.30	24.641	93
	$2 \rightarrow 1$	38.405	220.88	60.475	51
	Average	77.398	316.06	136.369	70.1
Static power (nW)	Holding "0"	62.969	216.61	17.406	11.953
	Holding "1"	2405.2	7667.9	4883.4	1331
	Holding "2"	101.98	206.35	16.792	12.838
Average power (μW)		1.1835	2.5143	1.4918	0.523

effect is still very difficult. Most of the research on these devices are still based on rudimentary and unreliable experimental prototypes or computational models. Industry-grade simulation tools and fabrication processes for these novel devices will take a long time to become reality.

4.5 Application of Ternary Sequential Circuits

Like the binary sequential circuits, the main applications of the ternary and higher-valued sequential circuits would be in designing shift registers, counters, encoders etc. Here, the basic design of a ternary shift register is briefly illustrated.

A shift register is formed by cascading several flip-flops in such a way that the output of one flip-flop is connected to the input of another flip-flop. A single shared clock signal results the stored data to shift from one register or flip-flop to another. By connecting the output of the last flip-flop to the input of the first flip-flop, the data can be stored for a long period of time, thereby using it as a form of computer memory. Shift register can be of different types: Serial-In-Serial-Out (SISO), Serial-In-Parallel-Out (SIPO), Parallel-In-Serial-Out (PISO) etc. The most basic of them is the Serial-In-Serial-Out (SISO) type. The basic structure of a ternary shift register is as shown in Fig. 4.9, where each single bit element is a basic ternary flip-flop and all of them are controlled by a single clock signal. The design and operation of this type of ternary shift register have been elaborated in [24, 25].

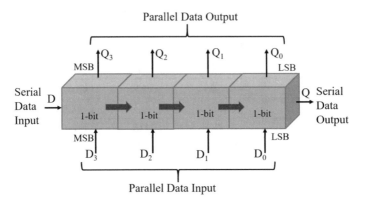

Fig. 4.9 4-bit shift register using D flip-flop [23]

The counter is a special type of shift register and mostly based on Serial-In-Parallel-Out (SIPO) type shift register. Here the output of the rightmost flip-flop is fed as the input of the leftmost flip-flop. As a result, the counter produces a certain sequence of data which is repeated every 'N' clock cycles. In [25], some ternary flip-flop-based counter and decoder designs are presented.

4.6   Conclusion

In this chapter, the fundamental idea of combinational and sequential circuit and the way to derive ternary sequential circuit from the concept of binary sequential circuit is presented. Here different approach for designing ternary latch, flip-flop etc. has been discussed. A basic power-delay comparison is also provided here in between different technologies and design approaches. Later, a sample implementation of the ternary sequential circuit is also discussed. Though, the concept of MVL sequential circuits, shifter, counter has been tried for some time, in modern world, this idea is yet to be apprehended.

References

1. Wu X, Prosser F. Ternary CMOS sequential circuits. In [1988] Proceedings. The Eighteenth International Symposium on Multiple-Valued Logic 1988 (pp. 307–313). IEEE.
2. Mouftah HT, Jordan IB. Design of ternary COS/MOS memory and sequential circuits. IEEE Transactions on Computers. 1977 Mar 1(3):281–8.
3. Jizhong S, Xiexiong C. Design of ternary flip-flops and sequential circuits based upon U_h gate. Journal of Electronics (China). 1993 Oct 1;10(4):356–64.

4. Lin S, Kim YB, Lombardi F. CNTFET-based design of ternary logic gates and arithmetic circuits. IEEE transactions on nanotechnology. 2009 Nov 24;10(2):217–25.

5. Moaiyeri MH, Mirzaee RF, Doostaregan A, Navi K, Hashemipour O. A universal method for designing low-power carbon nanotube FET-based multiple-valued logic circuits. IET Computers & Digital Techniques. 2013 Jul 1;7(4):167–81.

6. Sandhie ZT, Ahmed FU, Chowdhury MH. Design of Ternary Master-Slave D-Flip Flop using MOS-GNRFET. In 2020 IEEE 63rd International Midwest Symposium on Circuits and Systems (MWSCAS) 2020 Aug 9 (pp. 554–557). IEEE.

7. M. H. Moaiyeri, M. Nasiri, and N. Khastoo. (2016, Mar.). An efficient ternary serial adder based on carbon nanotube FETs. Engineering Science and Technology, an Int. J. 19(1), pp. 271–278.

8. M Hang, Guoqiang, and Xuanchang Zhou. "Novel CMOS ternary flip-flops using double pass-transistor logic." In 2011 International Conference on Electric Information and Control Engineering, pp. 5978–5981. IEEE, 2011.

9. Mirzaee, Reza Faghih, and Niloofar Farahani. "Design of a Ternary Edge Sensitive D FFF for Multiple-Valued Sequential Logic." Journal of Low Power Electronics 13, no. 1 (2017): 36–46

10. https://www.daenotes.com/electronics/digital-electronics/flip-flops-types-applications-woking.

11. Mouftah, H. T., and I. B. Jordan. "Design of ternary COS/MOS memory and sequential circuits." IEEE Transactions on Computers 3 (1977): 281–288.

12. A. P. Dhande and V. T. Ingole. (2005, Apr.). Design of 3-valued R-S & D flip-flops based on simple ternary gates. Int. J. Software Engineering and Knowledge Engineering, 15(2), pp. 411–417.

13. Sipos, E., and C. Miron. "Master-slave ternary D flip-flap-flops with triggered edges control." In 2010 IEEE International Conference on Automation, Quality and Testing, Robotics (AQTR), vol. 2, pp. 1–6. IEEE, 2010.

14. Zhou X, Hang G. Design of ternary D flip-flop using one latch with neuron-MOS literal circuit. In 2013 Ninth International Conference on Natural Computation (ICNC) 2013 Jul 23 (pp. 272–276). IEEE.

15. Wu, Xunwei, and Franklin Prosser. "Ternary CMOS sequential circuits." In [1988] Proceedings. The Eighteenth International Symposium on Multiple-Valued Logic, pp. 307–313. IEEE, 1988.

16. Madhuri, Badugu Divya, and S. Sunithamani. "Design of Ternary D-latch Using Graphene Nanoribbon Field Effect Transistor." In 2019 International Conference on Vision Towards Emerging Trends in Communication and Networking (ViTECoN), pp. 1–4. IEEE, 2019.

17. Qian, Wang, Wang Pengjun, and Gong Daohui. "Novel low-power ternary explicit pulsed JKL flip-flop based on CNFET." In 2016 IEEE International Conference on Ubiquitous Wireless Broadband (ICUWB), pp. 1–4. IEEE, 2016.

18. Jimmy, Sutaria, and Satish Narkhede. "Design of ternary D latch using carbon nanotube field effect transistors." In 2015 2nd International Conference on Electronics and Communication Systems (ICECS), pp. 151–154. IEEE, 2015.

19. Gope, Jayanta, Sanjay Bhadra, S. Chanda, M. Sarkar, S. Pal, and A. Rai. "Modelling of single electron ternary flip-flop using SIMON." In 2016 IEEE 7th Annual Ubiquitous Computing, Electronics & Mobile Communication Conference (UEMCON), pp. 1–9. IEEE, 2016.

20. Inaba, Motoi, Koichi Tanno, and Okihiko Ishizuka. "Multi-valued flip-flop with neuron-CMOS NMIN circuits." In Proceedings 32nd IEEE International Symposium on Multiple-Valued Logic, pp. 282–288. IEEE, 2002.

21. Uemura, Tetsuya, and Toshio Baba. "A three-valued D-flip-flop and shift register using multiple-junction surface tunnel transistors." IEEE Transactions on Electron Devices 49, no. 8 (2002): 1336–1340.

22. Lin M, Lü WF, Sun LL. Design of ternary D flip-flop with pre-set and pre-reset functions based on resonant tunneling diode literal circuit. Journal of Zhejiang University Science C. 2011 Jun 1;12(6):507–14.

23. ElectronicsTutorials. Retrieved from: https://www.electronics-tutorials.ws/sequential/seq_5. html

24. Amirany A, Jafari K, Moaiyeri MH. High-Performance Spintronic Nonvolatile Ternary Flip-Flop and Universal Shift Register. IEEE Transactions on Very Large Scale Integration (VLSI) Systems. 2021 Feb 12.

25. Dhande AP, Ingole VT, Ghiye VR. Ternary digital system: Concepts and applications. SM Online Publishers LLC; 2014 Oct 1.

MVL Random Access Memory

5

5.1   Introduction

Semiconductor memories can be classified into the following two categories based on access patterns:

1. Non-Random Access Memory: In this type of memories, the order of accessing the data for read or write operation is restricted. Shift registers, Content Addressable Memory (CAM) etc. fall under this category.
2. Random Access Memory: In random access memory, memory locations can be accessed for reading or writing of data in a random order. Most of the Read-Only-Memory (ROM), Read-Write-Memory (RWM) and Non-Volatile-Read-Write-Memory (NVRWM) fall into this class. However, in literature, the acronym RAM is reserved only for the random-access RWM.

In general, RAMs are volatile in nature. Volatile type of memory represents the group of memories where the stored data or information is lost when the power supply is turned off. RAMs can be design using static or dynamic circuit techniques leading to two types of RAMs: Static Random-Access Memory (SRAM) and Dynamic Random-Access Memory (DRAM). In a RAM, the time required for the reading or writing of a data is almost independent of the physical location inside the memory where the data is being read/written. On the other hand, in the direct-access data storage media (e.g., CD-RWs. DVD-RWs and older magnetic tapes and drum memory etc.), the time needed for the read and write operations changes considerably with the change of the physical location where the data is being read or written [1]. In current computing systems, RAM devices are designed using the conventional Metal–Oxide–Semiconductor Field Effect Transistor (MOSFET) technology. For beyond-CMOS platforms, other technologies are being explored for both binary and MVL RAMs. For ternary SRAM design, emerging technologies like Carbon Nano Tube Field Effect Transistor (CNTFET) [2–5], Fin Field Effect Transistor (Fin-FET) [6], Single Electron Transistor (SET) [7], SET-MOSFET hybrid [8], and Resonant Tunneling Diode (RTD) [9] are being explored along with the conventional MOSFET [8].

5.2   Static Random-Access Memory (SRAM)

Static Random-Access Memory (SRAM) is based on bi-stable latches or flip-flops, which are generally constructed by two inverters connected in a positive feedback loop. The standard binary SRAM cell contains six transistors. Four of these transistors constitute two cross-coupled inverters and the remaining two transistors act as the access devices that are used for reading and writing data of the SRAM cell. A single SRAM cell is often called an SRAM bitcell, which is the fundamental building block of an SRAM memory. Each SRAM cell has two bitlines (BL) that contain the stored bit and its' complement. Another signal line known as the wordline (WL) controls the access transistors. SRAM cells do not require any refreshing circuit and as a result the operation of an SRAM cell is simple. Another advantage of the SRAM cell is that ideally it does not consume power during the idle mode of operation. However, higher number of transistors per memory cell results in higher area and power consumption, higher parasitic values and lower memory density [10]. In addition to the standard binary 6T SRAM cell, there are some designs with lower or higher number of transistors per cell. Some of these designs are known as the 4T, 8T, and 10T SRAM [11–13].

In case of ternary SRAMs, most of the designs adopt structures like the binary SRAMs. The major differences between the ternary and the binary SRAMs are the differences in the designs of the ternary inverter and the access transistor. Many ternary SRAM designs have been proposed and most of these designs require higher number of transistors compared to the binary counterparts. Additional transistors are required to ensure a stable voltage transfer curve for the standard ternary inverter (STI), which also contains some resistors. Use of physical resistors in any design leads to higher area consumption and parasitic effects. More recent designs focus on avoiding the use of physical resistors by adding additional transistors to obtain a stable transfer curve, which in turn results in a lower noise margin. In the next two subsections, two basic ternary SRAM designs are illustrated.

5.2.1   Design 1 of a Ternary SRAM

The schematic diagram of one of the ternary SRAM designs is shown in Fig. 5.1 [4]. This design is very similar to the standard binary SRAM. Instead of using the conventional

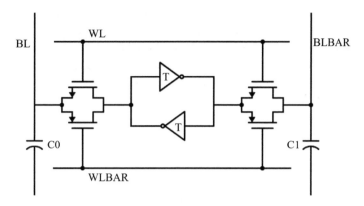

Fig. 5.1 The first design of a ternary SRAM based on CNTFET [4]

MOSFET, the proposed design uses CNTFET technology to implement the ternary SRAM circuit [4].

In this ternary SRAM design, the binary inverters of the positive feedback loop of the SRAM cell are replaced by 5T ternary inverters. Transmission gates are used as the access transistors to ensure correct read/write operation in the memory cell without any threshold voltage drop. The access transistors are gated to the wordlines WL and WLBAR. The data and the complimented form of the data are fed through the bitlines BL and BLBAR. The write and read operations of this SRAM cell are discussed below.

Write Operation

During the write operation, the data that needs to be written in the memory cell is placed on the bitlines BL and BLBAR. Once the wordlines WL and WLBAR are asserted, the access transistors are turned on and the data is copied to the nodes Q and QBAR. After the wordlines are de-asserted, the data stored at the nodes Q and QBAR are held by the back-to-back inverter loop.

Read Operation

During the read operation, the bit lines BL and BLBAR are pre-charged to a value of $\frac{1}{2}V_{DD}$. As soon as the wordlines WL and WLBAR are asserted, the access transistors are turned on. The data that were stored in Q and QBAR are passed to the bitlines and the pre-charge capacitors' values are altered. When the node Q and QBAR are at logic 2 and 0 respectively, the pre-charged capacitor C0 is pulled up to V_{DD} and C1 is pulled down to ground once the wordlines are asserted. Again, if the node Q and QBAR is at logic 0 and 2 respectively, the pre-charged capacitor C0 is pulled down to ground and C1 is pulled up to V_{DD}. If Q and QBAR are holding the logic 1, then the pre-charge capacitors remain at the previous value. A sense amplifier is used to sense the difference between the two capacitors and make an informed decision about the stored logic level. The sizing of the transistors must be determined carefully to avoid read upset.

Another similar ternary SRAM design is shown in [8], where MOSFET technology is used. Here, each ternary inverter is designed using 4 MOSFET transistors. The operating conditions of this MOSFET based design are slightly different from the CNTFET based design explained in this subsection. However, from overall structural point of view, both ternary SRAM designs are quite similar.

5.2.2 Design 2 of a Ternary SRAM

Another design approach for ternary SRAM (as shown in Fig. 5.2) were tried by some researchers [2, 3]. This design was also implemented using CNTFET. Two ternary inverters connected back-to-back in a positive feedback loop constitute the storage element. Two transmission gates are used as the read and write access devices. In the design shown in Fig. 5.2, there is an additional read buffer (compared to Fig. 5.1) to ensure reading of correct data from the memory cell. The read buffer is made of one P-type CNTFET and one N-type CNTFET. The threshold voltages of these two transistors are −0.559 V and 0.559 V respectively. The purpose of the read buffer is to invert the logic state "0" and "2" stored in the node QB. It also prevents the read upset problem and increases the static noise margin of the memory cell. The design shown in Fig. 5.2 has two sets of wordlines—one pair for the "write" operation and the second pair for the "read" operation.

The read and write operation of the ternary SRAM cell illustrated in Fig. 5.2 are explained below.

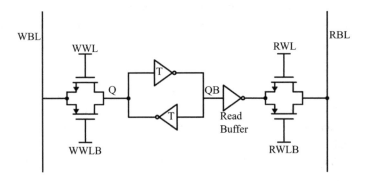

Fig. 5.2 The second design of a ternary SRAM based on CNTFET [2, 3]

Write Operation

The purpose of the write transmission gate is to write the correct data into the inverter-loop. When the write wordline WWL is high and the WWLB is low, the write transmission gate turns on. As a result, the data held in the write bitline WBL is written in the memory cell.

Read Operation

For the read operation, the read bitline RBL is pre-charged to ½V$_{DD}$. During the read operation both the read buffer and the read transmission gate are involved. When the data stored in the cell is logic "0", the node Q remains at 0 V and QB remains at V$_{DD}$. The read buffer inverts the voltage at node QB. As a result, the output at the read buffer becomes 0 V. As soon as the read wordline RWL becomes high and RWLB becomes low, the read transmission gate turns on. After that, the read bitline RBL is discharged to logic "0" through the transmission gate and the read buffer. If the data stored in the SRAM cell is "2", the node Q remains at V$_{DD}$ and QB remains at 0 V. The value of QB is inverted through the read buffer and becomes logic "2". Thus, the read bitline assumes the logic "2" when the read word lines RWL and RWLB are accessed. If the data stored in the cell is logic "1", both node Q and QB remains at logic "1". The read buffer passes a value "0" as both the transistors of read buffer are turned off. In that case, the read bitline RWL is kept at the pre-charged voltage once the read bitlines are asserted.

5.2.3 Analysis

In addition to the CNTFET based ternary SRAM designs shown in Figs. 5.1 and 5.2, other technologies like FinFET were also explored for ternary SRAM. Table 5.1 provides a comparative analysis of different ternary SRAM designs. The simulation results for the conventional CMOS based SRAM cell are taken from [2]. The FinFET based design presented in [6] follows the binary SRAM convention for ternary inverter design and uses two transistors to implement a ternary inverter. Most of the other designs use the 6T ternary inverter implementation to achieve higher noise margin [14]. Though using 2T-based ternary inverter seems more efficient and cost-effective than the 6T-based ternary inverter, in reality, the logic behind using multiple transistor while designing ternary inverter is obtaining a stable voltage transfer curve.

5.3 Dynamic Random-Access Memory (DRAM)

Dynamic Random-Access Memory (DRAM) is designed using the dynamic memory circuit technique where the logic voltages are held by the capacitive nodes instead of having a positive feedback loop. Like the SRAM the data stored in the DRAM is also lost once

Table 5.1 Quantitative analysis of different ternary SRAM design

Technology		Conventional CMOS SRAM cell [2]	CNTFET (18 T) [2]	CNTFET (18 T) [3]	CNTFET (14 T) [4]	CNTFET (10 T) [5]	FinFET (6 T) [6]
Write delay (ps)	0 to >1	–	–	113.1	2.18	28.2	–
	1 to >2	–	3.71	5.85	2.7	4.27	–
	2 to >0	–	15.66	10.9	3.61	4.11	–
	0 to >2	–	6.28	20.64	3.26	2.43	–
	2 to >1	–	22.24	95.91	2.61	39.8	–
	1 to >0	286.23	9.06	7.19	3.42	3.59	–
	Average	286.23	11.39	42.27	2.96	13.7	3.825
Read delay (ps)	Read 0	110.8	55.16	127.6	8.12	43.5	–
	Read 1	100.2	18.9	39	1.02	15.2	–
	Read 2	–	20.14	135.5	9.44	57.3	–
	Average	105.5	31.4	100.7	6.2	38.67	10.165
Write power (uW)	Logic 0	–	–	–	2.73	–	–
	Logic 1	–	–	–	0.17	–	–
	Logic 2	–	–	–	2.74	–	–
	Average	0.245	0.2	–	1.88	–	680
Read power (uW)	Logic 0	–	–	–	3.6	–	–
	Logic 1	–	–	–	4.16	–	–
	Logic 2	–	–	–	3.5	–	–
	Average	0.57	0.15	-	3.75	–	935
Standby power (nW)	Storing 0	35.36	0.032	0.35	–	1470	–
	Storing 1	–	442	828.7	–	1350	–
	Storing 2	–	0.014	0.51	–	900	–

the power is turned off, which makes the DRAM a volatile device. A typical DRAM cell contains only a single transistor and a capacitor implemented by the standard MOSFET technology. The integrity of the data stored in the DRAM cell depends on the values of the voltage or the amount of the charge held by the capacitor of the logic nodes. Depending on the charged or discharged state of the logic node, the value of the data can be assessed as "0" or "1" in case of binary DRAM. Since all capacitors tend to discharge from the charged condition, the stored data is distorted with time. Therefore, DRAM needs an additional refreshing circuit to periodically restore the data to its original state. Addition of the refreshing circuit makes DRAM design and operation more complicated compared to the SRAMs. However, the structural simplicity of a single-transistor DRAM cell results in a denser memory chip which is turn reduces the chip area and cost of each DRAM cell [10]. As the operation and the integrity of DRAM cells depend on the charging and discharging of parasitic capacitors, the design of ternary or quaternary DRAM cells are significantly more complicated than binary DRAM or ternary/quaternary SRAM. In the next subsection, a design of a ternary DRAM is illustrated.

5.3.1 Design 1 a Ternary DRAM

The design proposed in [15] is most similar to the binary 3T DRAM. The schematic diagram of the design is given in Fig. 5.3. The design is based on CNTFET and uses a single wordline. Here, transistor M_1 and M_2 are N-type and M_3 is P-type in nature. One word-line (WL) and two bit-lines (BL) are being used for the appropriate operation of the memory cell. Word line WL controls the read and write operation. If WL is high, the write operation takes place and is WL is low, then read operation takes place. The bit lines are used for read and write operation directly (BL_1 for write and BL_2 for read). The threshold voltages for M_1, M_2, and M_3 are 0.24 V, 0.6 V and -0.24 V respectively. Two capacitors are used for the proper functioning of the memory cell namely CS and CBL.

Write Operation
Bit line BL_1 contains the data that is required to be written to the memory cell. If the word line WL becomes high, the transistor M_1 becomes on. Therefore, the capacitor CS charges according to the data available in BL_1. If BL_1 contains logic 1 or $1/2V_{DD}$, then the same value gets stored in the capacitor CS as well as node X. If BL_1 contains logic 2

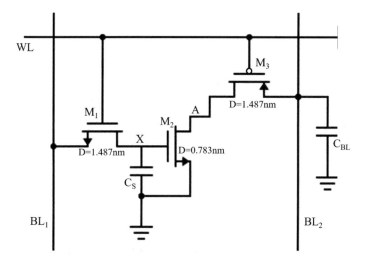

Fig. 5.3 The first design of ternary 3T-Dram cell with single wordline [15]

then a threshold voltage drop (V_{th}) occurs and a voltage equal to V_{DD}-V_{th} is obtained at the node X. This data is held by the node X until the next positive cycle of WL.

Read Operation
The read operation takes place when WL becomes equal to 0 V. During that time, the transistor M_3 turns on. The bit line BL_2 is pre-charged to a voltage equal to V_{DD}. If the data held by the node X is "0", then transistor M_2 is turned off. As a result, the voltage at node A remains high. When the data stored in the memory cell is "1" or "2", then transistor M_2 turns on. So, the voltage at bit line BL_2 (which is stored by the capacitor CBL) starts to discharge through M_2. The rate at which the bit line is discharged denotes the value that was being held by the memory cell. The read operation in inverting in nature.

5.3.2 Design 2 a Ternary DRAM

The design proposed in [16] uses 5 transistors for the implementation of a ternary DRAM cell (shown in Fig. 5.4). The design is comprised of one write transistor (PW), two feedback transistors (PB and PF), and two read transistors (PR and PA). Compared to a basic 3T DRAM cell, the ternary DRAM cell design presented in [16] are different in two ways. The first difference is the use of all PMOS transistors in place of NMOS transistors and the second difference is the use of two extra transistors for an internal feedback mechanism. These feedback transistors ensure robust data storage and limit the power required to refresh the memory cell.

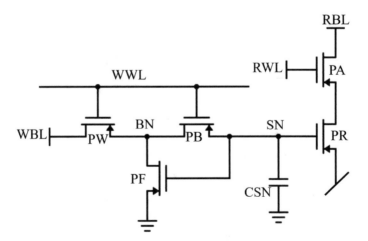

Fig. 5.4 The second design of ternary 5-T DRAM cell based on ternary gain-cell eDRAM [16]

The read and write operation of the ternary DRAM cell shown in Fig. 5.4 are explained below.

Write Operation

The data that needs to be written in the memory cell are placed in the write bitline WBL. Write operation takes place as soon as the word bitline WWL is set to a very low voltage (0.7 V). At that time, the PW and PB are turned on and pass the value on WBL to the parasitic capacitor CSN. As a result, the appropriate data is written on the node SN.

Read Operation

For the read operation, the read bitline RBL is pre-discharged to 0 V. And the word bitline WBL is pre-charged to V_{DD}. The read operation starts as soon as the WBL becomes 0 V (logic 0). At that time, the read bitline RBL changes the charged state depending on the stored value at the memory cell. If the stored value at the cell is 0, RBL is charged to V_{DD} after a short time. If the stored value at the cell is logic 1, RBL starts to change its' value at a slower pace. And if the stored value at the cell is logic 2, RBL remains at the previous value of 0 V. A sense amplifier is used to sense the change of voltage level and the speed of the change of voltage level.

5.3.3 Analysis

Common DRAM cells are 3-transistor (3T) and 1-transistor (1T) based designs. In a 3T DRAM cell, two capacitors are required to ensure proper functioning of the memory cell. In a 1T DRAM cell, one capacitor is required. The purpose of these capacitors is to store a certain amount of charge corresponding to the signals' logical values. The switching transistors transmit the charge between the cell and the bitlines. In a conventional binary DRAM, the signal charge is half (1/2) of the maximum stored charge. Whereas, in a quaternary memory element, the signal charge is one-sixth (1/6) of the maximum stored charge [17]. As a result, the value of the capacitor in a quaternary DRAM cell needs to be three times larger than the binary DRAM [18]. Typically, higher density in DRAM memory designs is achieved by sacrificing some design rules and improving cell structure. Multiple-valued storage technique offers a huge increase of storage density in DRAM memory. The first 4 gigabit 4-level (quaternary) DRAM in the world was developed by NEC [19].

It is possible to convert any 1T or 3T binary DRAM cell into a ternary or quaternary DRAM cell by setting different intermediate voltage levels in between the ground and the supply voltage. The main challenge with this approach is the interpretation of the stated voltage levels using sense amplifiers or some other special circuitry. For MVL based DRAM design, a limited number of research efforts to use conventional and emerging technologies can be found in the literature. Couple of examples are MOSFET based MVL DRAM design presented in [16] and the hybrid SET-MOSFET based design presented in [20].

5.4   Multi-level Dynamic Random-Access Memory (MLDRAM)

If more than two logic levels are stored in a single DRAM cell, the noise margin of the memory cell decreases significantly. The Multi-Level DRAM (MLDRAM) is based on the concept that the DRAM cell itself would not handle multiple voltage levels independently. Rather, a set of associated circuitries will be utilized along with the DRAM cell to achieve multiple voltage levels using voltage divider rule. Majority of the researchers who are working on MLDRAM focus on holding four different logic levels by obtaining 2 bits in a single cell (in contrast, in binary design, a single cell can give 1 bit at a time). For storing four logic values, the circuit must be capable of producing seven voltage levels—V_{DD}, V_{SS} (ground), and 5 intermediate voltage levels as shown in Fig. 5.4 [21]. As a result, complicated sensing circuit and accurate reference voltage generation become major concerns. To deal with these concerns, various serial and parallel sensing circuitry, different charge sharing techniques etc. have been proposed [22–25].

The state flow diagram of an MLDRAM is shown in Fig. 5.5 [26]. In this case, the supply voltage is considered to be 1.8 V. According to this diagram, the read cycle compares the stored data with a reference voltage of $\frac{1}{2}V_{DD}$. After that, based on the information gained from the sense amplifier, one or zero is assigned to the Most Significant Bit (MSB). The X value inside the circle stands for unknown Least Significant Bit (LSB). After that, the stored value is compared with the LSB voltage reference.

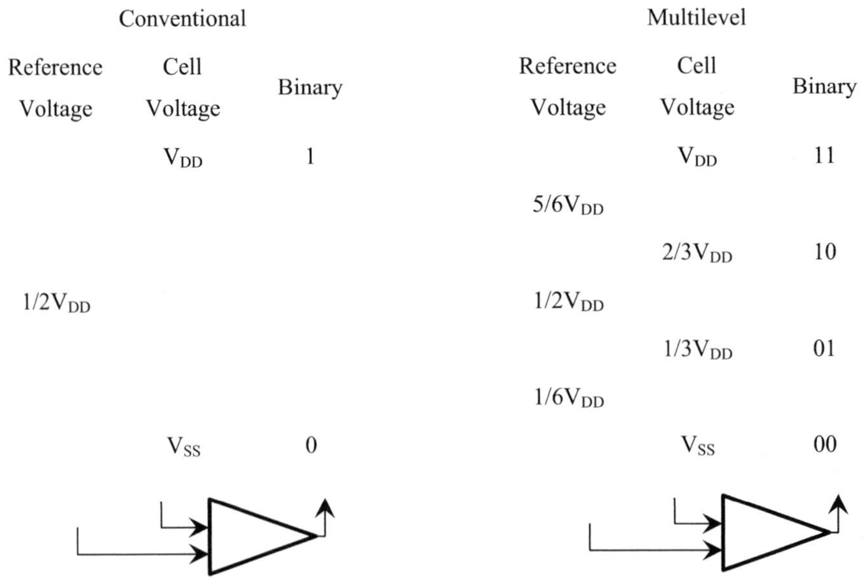

Conventional				Multilevel		
Reference Voltage	Cell Voltage	Binary		Reference Voltage	Cell Voltage	Binary
	V_{DD}	1			V_{DD}	11
				$5/6V_{DD}$		
					$2/3V_{DD}$	10
$1/2V_{DD}$				$1/2V_{DD}$		
					$1/3V_{DD}$	01
				$1/6V_{DD}$		
	V_{SS}	0			V_{SS}	00

Fig. 5.5 Storage diagram for a 2 bits per cell design [21]

Generally, 1T1C (1 Transistor 1 Capacitor) binary DRAM cell is used to the design the MLDRAM. As mentioned earlier, the DRAM cell itself is not independently responsible for providing multiple voltage levels. The associated circuitry obtains multiple voltage levels using voltage divider rule. Many MLDRAM designs have been proposed in recent time. Some of the notable MLDRAM design examples are Okuda and Murotani [22], Furuyama et al. [23], Gillingham [24], and Birk et al. [25]. A detailed discussion of these designs is available in [25]. In the next subsection, the MLDRAM design proposed by Gillingham [24] is briefly illustrated.

5.4.1 MLDRAM Design Proposed by Gillingham

The schematic diagram of the proposed MLDRAM by Gillingham [24] is shown in Fig. 5.6 [27]. The circuit can be visualized as two sub-bit line pairs (left and right) which interact through a switch matrix comprised of the signals C, CBAR, X, and XBAR. The left (right) sub-bit line pair is comprised of BL(BR) and BLBAR (BRBAR), which are connected to the inverting and non-inverting terminals of the left (right) sense amplifier, respectively. Each cell of an array contains a capacitor, which is connected in between a common biased plate node and the source node of an access transistor. The drain of each access transistor is connected to a sub-bit line. During the read operation of a cell in the left (right) of the cell array, the MSB can be recovered from the left (right) sense amplifier and the LSB can be recovered from the right (left) sense amplifier. Here, the precharge voltage $V_{pre} = 1/2\ V_{DD}$.

Regarding an addressed cell in the left or right array, the MSB side is referred to the array that contains the cell and the LSB side refers to the other side of the switch matrix. The word lines are numbered as WL_1, ..., WL_{i+1} and WR_1, ..., WR_{i+1} for the left and right cell arrays. The operation is done on the cell denoted by WL_i, which is located at the left complimentary bit line. Dummy word lines are connected as a precaution for the reduced noise margin of the MLDRAM. Dummy word line DLo (DRo) is addressed during the access of the cells which are connected to the odd-numbered word lines in the left (right) array. Similarly, dummy word line DLe (DRe) are addressed during the access of the cells which are connected to the even-numbered word lines. Figure 5.6 shows the additional control signals that take part in the proper functioning of the memory cell. The signals IL and IR are used to control the pass transistors that allows the cell array to be separated from the sense amplifiers. The equalizer signals EL and ER can be used to independently short the sub-bit line pairs. And the pre-charge signals PL and PR can be used to independently pre-charge the sub-bit lines to a voltage of $1/2V_{DD}$. ZL and ZR acts as similar equalizer and pre-charge control signal for the sense amplifiers. YSL and YSR are used as connection between left and right sub-bit line pairs and differential data buses. The control waveform of a MLDRAM is shown in Fig. 5.7 [27] (Fig. 5.8).

The operation of an MLDRAM can be divided into three phases as explained below.

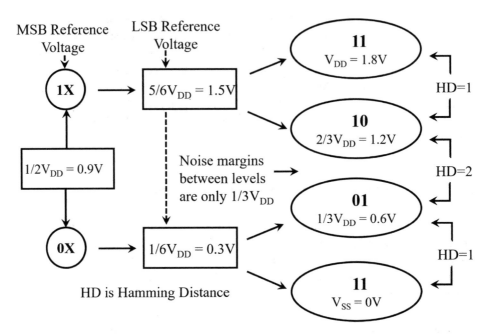

Fig. 5.6 Basic operating diagram for an MLDRAM [26]

Write Operation
When the control signals YSL and YSR are asserted the MSB and the LSB values are
passed to the sub-bitlines from the data bus. Then the control signal IR becomes low,
which in turn disconnects the right bitline from the sense amplifier. As a result, the LSB
is captured by the sub-bitline BRBAR. After that, XBAR is asserted, which connects
BLBAR and BR and both bitlines carry the MSB. At that time, left sense amplifier is dis-
connected from the circuit and the control signal ER is pulled up, which allows BRBAR,
BLBAR and BR to hold the total charge. The resultant signal voltage, encoding both the
LSB and MSB is (2/3MSB + 1/3LSB) * V_{DD}.

Isolate and Store Operation
During this phase the word signal WLi is asserted, and the dummy word signals are de-
asserted. This causes the capture of the stored value into the cell. Then all the sub-bitlines
are pre-charged to the value $1/2V_{DD}$. The sense amplifiers are then turned off and the
outputs are pre-charged to $1/2V_{DD}$.

Read Operation
During the read operation the signals WLi and CBAR are asserted sequentially. As a
result, the signal stored in the cell is divided in between the sub-bitlines BLBAR and

Fig. 5.7 Schematic diagram of the MLDRAM design proposed by Gillingham [24, 27]

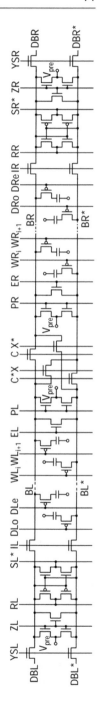

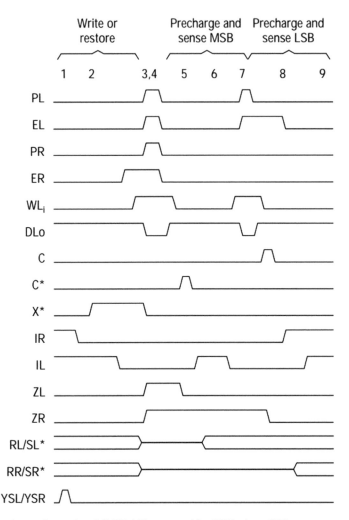

Fig. 5.8 Control waveform of an MLDRAM proposed by Gillingham [27]

BRBAR. The comparison between the stored values on BLBAR and BL (which are pre-charged to $1/2V_{DD}$) is done by the sense amplifier. The amplification of the signal to a full-swing differential MSB is done by the sense amplifier, which is applied to the BL and BLBAR. Then the control signal IL is de-asserted and the recovered MSB value is latched to the left sense amplifier. The wordline WLi is then asserted, which captures the full-swing MSB signal to the cell. As a result, the reference voltage LSB is created for the BR. Next, the left bitlines are equalized and pre-charged to $1/2V_{DD}$. Afterward, the control signals EL and C go low and the MSB signal is transferred to the bitlines BLBAR, BL and BR. The resulting reference signal of BR is then isolated. Later, the control signal IR goes high and the comparison between the stored voltage on BRBAR and the reference voltage on BR is done by the right sense amplifier. The amplification of the signal to a full-swing differential LSB is done by the sense amplifier which is applied to the BR and BRBAR. Then obtained value is amplified up to the full-swing voltage and data is ready to be propagated to the data bus. The read operation of an MLDRAM is self-destructive like conventional DRAM. Therefore, a restore operation is needed, which is like the write operation.

5.4.2   Analysis

An MLDRAM offers increased information density where more than two logic bits can be stored simultaneously in a single storage cell. However, this benefit comes at a price of reduced speed, increased latency, complicated circuitry, and reduced reliability in terms of soft error rate. As of now, the interest in MLDRAM has been mostly academic. As an attempt to increase the reliability of MLDRAM, different Error Correcting Codes (ECC) have been proposed. Unfortunately, the conventional ECC circuit does not work well with the MLDRAM [27]. Therefore, dedicated ECC units must be developed for MLDRAM [26, 27].

5.5   Conclusion

This chapter provides an overview of Random-Access Memory (RAM). The basic design and operation of Static RAM (SRAM) and Dynamic RAM (DRAM) are explained. Ways to modify basic binary SRAM and DRAM cells for designing Multiple Valued Logic (MVL) based SRAM and DRAM are illustrated. Several multi-level SRAM and DRAM designs proposed by different researchers are investigated. The proposed designs explore both the conventional (MOSFET) and the emerging device technologies (CNTFET, SET, FinFET, etc.) to implement the multi-valued RAM cells. Considering the critical role RAM devices play in all computing and micro and nano-electronic applications, this chapter is intended to provide a foundational understanding of MVL RAM.

References

1. Course Hero "Introduction to Computer Applications and Concepts Module 2: Computer Hardware", Retrieved from: https://www.coursehero.com/study-guides/zeliite115/reading-random-access-memory/, Retrieved on: 25th May, 2022.
2. Shreya S, Sourav S, "Design, analysis and comparison between CNTFET based ternary SRAM cell and PCRAM cell" in Communication, Control and Intelligent Systems (CCIS), pp. 347–351, 2015.

3. Lin S, Kim YB, Lombardi F. Design of a ternary memory cell using CNTFETs. IEEE Transactions on Nanotechnology. 2012 Aug 3;11(5):1019–25.
4. Niranjan, S., Shanmukha Sandesh, and Vasundara Patel KS. "A Novel Design of Ternary Level SRAM cell using CNTFET." In 2018 International Conference on Networking, Embedded and Wireless Systems (ICNEWS), pp. 1–5. IEEE, 2018.
5. Srinivasan, Pramod, Anirudha S. Bhat, Sneh Lata Murotiya, and Anu Gupta. "Design and performance evaluation of a low transistor ternary CNTFET SRAM cell." In 2015 International Conference on Electronic Design, Computer Networks & Automated Verification (EDCAV), pp. 38–43. IEEE, 2015.
6. Kishor, Makani Nailesh, and Satish S. Narkhede. "Design of a ternary FinFET SRAM cell." In 2016 Symposium on Colossal Data Analysis and Networking (CDAN), pp. 1–5. IEEE, 2016.
7. Syed, Naila, and Chunhong Chen. "Low-power multiple-valued SRAM logic cells using single-electron devices." In 2012 12th IEEE International Conference on Nanotechnology (IEEE-NANO), pp. 1–4. IEEE, 2012.
8. Wanjari NP, Hajare SP. VLSI design and implementation of ternary logic gates and ternary SRAM cell. International Journal of Electronics and Computer Science Engineering. 2013;2(2):610–8.
9. Wei, S-J., and Hung Chang Lin. "A multi-state memory using resonant tunneling diode pair." In 1991, IEEE International Symposium on Circuits and Systems, pp. 2924–2927. IEEE, 1991.
10. https://www.diffen.com/difference/Dynamic_random-access_memory_vs_Static_random-acc ess_memory.
11. Kulkarni, Jaydeep P., Keejong Kim, and Kaushik Roy. "A 160 mV robust Schmitt trigger based subthreshold SRAM." IEEE Journal of Solid-State Circuits 42, no. 10 (2007): 2303–2313.
12. Kosonocky, Stephen V., and Azeez Bhavnagarwala. "Quasi-static random access memory." U.S. Patent 6,975,532 issued December 13, 2005.
13. Mohammed, Mahmood Uddin, Nahid M. Hossain, and Masud H. Chowdhury. "A Disturb Free Read Port 8T SRAM Bitcell Circuit Design with Virtual Ground Scheme." 2018 IEEE 61st International Midwest Symposium on Circuits and Systems (MWSCAS). IEEE, 2018.
14. Sandhie, Z.T., Ahmed, F.U. and Chowdhury, M., 2020, February. GNRFET based Ternary Logic–Prospects and Potential Implementation. In 2020 IEEE 11th Latin American Symposium on Circuits & Systems (LASCAS) (pp. 1–4). IEEE.
15. Sandhie ZT, Ahmed FU, Chowdhury MH. Design of novel 3T ternary DRAM with single word-line using CNTFET. Microelectronics Journal. 2022 Jun 25:105498.
16. Maltabashi, Or, Hanan Marinberg, Robert Giterman, and Adam Teman. "A 5-Transistor Ternary Gain-Cell eDRAM with Parallel Sensing." In 2018 IEEE International Symposium on Circuits and Systems (ISCAS), pp. 1–5. IEEE, 2018.
17. T. Okuda, "Advanced circuit technology to realize post giga-bit DRAM," Proceedings. 1998 28th IEEE International Symposium on Multiple- Valued Logic (Cat. No. 98CB36138), 1998, pp. 2–5, doi: https://doi.org/10.1109/ISMVL.1998.679266.
18. Dubrova E. Multiple-valued logic in VLSI: challenges and opportunities. In Proceedings of NORCHIP 1999 Nov (Vol. 99, No. 1999, pp. 340–350).
19. NEC: NEC develops the world's first 4 gigabit DRAM. Retrieved from: http://www.nec.co.jp/press/en/9702/0602.html
20. Kye, H. W., S. J. Shin, T. H. Lee, and J. B. Choi. "One electron-controlled multi-valued dynamic random-access-memory." In 2016 IEEE Nanotechnology Materials and Devices Conference (NMDC), pp. 1–2. IEEE, 2016.
21. Bo Liu, J. F. Frenzel and R. B. Wells, "A multi-level DRAM with fast read and low power consumption," 2005 IEEE Workshop on Microelectronics and Electron Devices, 2005. WMED '05, 2005, pp. 59–62, doi: https://doi.org/10.1109/WMED.2005.1431619.

22. T. Okuda and T. Murotani, "Four-level storage 4-Gb DRAM," IEEE J. Solid-States Circuits, vol. 32, no. 11, pp. 1743–1747, Nov. 1997.
23. T. Furuyama, T. Ohsawa, Y. Nagahama, H. Tanaka, Y. Wanatabe, T. Kimura, K. Muraoka, and K. Natori, "An experimental 2-bit/cell storage DRAM for macrocell or memory-on-logic application," IEEE J. Solid-State Circuits, vol. 24, no. 2, pp. 388–393, Apr. 1989.
24. P. Gillingham, "A sense and restore technique for multilevel DRAM," IEEE Trans. Circuits Syst. II, Exp. Briefs, vol. 43, no. 7, pp. 483–486, Jul. 1996.
25. Birk G, Elliott DG, Cockburn BF. A comparative simulation study of four multilevel DRAMs. In Records of the 1999 IEEE International Workshop on Memory Technology, Design and Testing 1999 Aug 9 (pp. 102–109). IEEE.
26. Polianskikh B, Zilic Z. Induced error-correcting code for 2 bit-per-cell multi-level DRAM. In Proceedings of the 44th IEEE 2001 Midwest Symposium on Circuits and Systems. MWSCAS 2001 (Cat. No. 01CH37257) 2001 Aug 14 (Vol. 1, pp. 352–355). IEEE.
27. Redeker M, Cockburn BF, Elliott DG. Fault models and tests for a 2-bit-per-cell MLDRAM. IEEE Design & Test of Computers. 1999 Jan;16(1):22–31.
28. Yu, Y. S., H. W. Kye, B. N. Song, S-J. Kim, and J-B. Choi. "Multi-valued static random access memory (SRAM) cell with single-electron and MOSFET hybrid circuit.' Electronics Letters 41, no. 24 (2005): 1316–1317.

MVL Flash Memory

6

6.1 Introduction

Flash memory or Flash storage is a type of memory device which can be electronically erased and reprogrammed. It is non-volatile in nature, which means the value stored in a flash memory does not get lost upon turning OFF the device. It is one kind of Electronically Erasable Programmable Read Only Memory (EEPROM) that can be erased and reprogrammed in large blocks [1]. Flash memory is used for easy and fast information storage in computers, SD cards, mobile phone, digital cameras, video games, scientific instrumentation, industrial robotics, and medical electronics.

6.2 Floating Gate MOS (FGMOS)

The fundamental building block of flash memory is the floating-gate MOSFET (FGMOS), which was developed by Dawon Kahng and Simon Min Sze as a variation of the MOSFET to be used in Programmable-Read-Only-Memory (PROM) type memory cell. FGMOS is non-volatile and re-programmable in nature [2].

The structure of a FGMOS (shown in Fig. 6.1) is like a typical MOSFET except for an additional floating gate in between the channel and the gate. The floating gate is made of an extra polysilicon strip buried inside the insulation layer in between the gate and the channel. This extra gate made of polysilicon strip is not connected to any node and that is why it is called a floating gate. The two gates of the FGMOS device are identified as the control gate (CG) and the floating gate (FG). In the absence of any charge on the floating gate, the device operation is almost similar to a normal MOSFET. If a positive charge is applied to the control gate, then a channel is created between the source and drain and current flows from source to drain. Due to the insertion of this extra gate, the oxide

© The Author(s), under exclusive license to Springer Nature Switzerland AG 2022
Z. T. Sandhie et al., *Beyond Binary Memory Circuits*, Synthesis Lectures
on Digital Circuits & Systems, https://doi.org/10.1007/978-3-031-16195-7_6

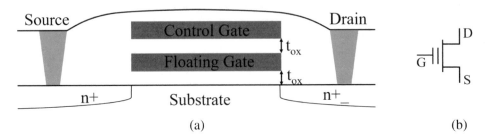

Fig. 6.1 **a** Structural diagram and **b** logic symbol of a floating-gate MOS (FGMOS)

thickness and the threshold voltage of the transistor increases. It is possible to program
the threshold voltage of the device by trapping of electric charge on the floating gate.

When a high enough voltage is applied to the gate and drain, a high electric field is created. An avalanche injection of electron takes place through the insulation layer between
the channel and the floating gate. This avalanche injection can happen through an oxide
layer of thickness up to 100 nm, which makes the fabrication process easy. These injected
electrons get trapped in the floating gate and decreases the effective voltage of the floating
gate. For this reason, this structure is often called Floating-Gate Avalanche-Injection MOS
or FAMOS. Upon the removal of the applied high voltage, the negative charge causes an
induced negative voltage at the floating gate. This negative charge increases the effective
threshold voltage of the device. As a result, a high voltage is required for turning the
device on. As the floating gate is insulated all around, the trapped charge remains stored
in the floating for a long time, even after the supply voltage is eliminated, thereby making
a nonvolatile memory cell.

If the threshold voltage of the device in the absence of a negative charge in the floating
gate is Vth_1, then the increased threshold voltage in the presence of a charge on the floating gate is Vth_2 ($Vth_2 > Vth_1$). If in a scenario in which the floating gate was previously
uncharged and an intermediated voltage between Vth_1 and Vth_2 is applied to the device,
then the device will start conducting, i.e. "1" is stored to the device. On the contrary, if
the FG is charged and an intermediate voltage is applied, the device will not conduct,
i.e. "0" is stored to the device. Therefore, the flow of a current upon the application of
an intermediate voltage would denote either "0" or "1" as the stored value in the device
depending on whether the floating is previously uncharged or charged. In case of a multi-
level cell system, not only the presence of current flow, but also the amount of current
flow is measured to estimate the amount of charge stored in the FG. The different levels
of current would denote different levels of logic that can be handled by a single cell.

6.3   ETOX Flash

ETOX device is an improvement of the original FGMOS device. The difference is in the use of a very thin oxide layer in the ETOX device between the floating gate and the channel. During the programming stage a high voltage is applied to the gate and the drain terminals with the source grounded. And during the erasure stage a high voltage is applied to the source with the gate grounded.

The initial versions of the floating-gate-transistor based memories include Erasable Programmable Read Only Memory (EPROM) and Electrically Erasable Programmable Read Only Memory (EEPROM). In these early versions of the nonvolatile memories only one bit of data could be handled at a time. In 1980, Fujio Masuoka, who was working for Toshiba, proposed a kind of floating-gate memory, where a complete section of memory could be erased and programmed simultaneously by applying a supply voltage to a particular wire connected to many cells [3]. Due to the capability of erasing and programming an entire block of memory cells at a time, the new nonvolatile memory was named as "Flash Memory".

6.4   NAND and NOR Flash Memory

Flash memories are grouped into two main types: NAND and NOR flash. In NOR Flash, one terminal of each memory cell is directly connected to the ground and the other terminal is connected to the bit line. The operation of the complete arrangement resembles a NOR gate, thereby it is called a NOR flash. In NAND flash, several memory cells (generally eight cells) are connected in series connection, like a NAND logic gate.

In a NOR flash memory, the transistors corresponding to the memory cells appear in parallel connection. As a result, it is possible to read and write data by activating each transistor individually in the NOR type flash memory. Therefore, NOR flash is used in applications where random access of data and execute-in-place access are needed. In NOR flash, enough addresses can be provided for the mapping of all memory range. Faster random access and less read time can be obtained, which makes it suitable for code execution. The drawback is low programming and erasing speed. If the memory cell is large enough, it becomes expensive [4].

NAND flash cell is smaller in size, more cost effective, and offers higher programming and erasing speed. On the other hand, it has lower reading speed, and it does not permit random access of the data. The code execution is different and more complicated [4]. Another disadvantage is the possibility of the presence of bad blocks. It requires some Error Correcting Code (ECC) function within the device.

The schematic diagrams of a NOR and a NAND flash memory cell are shown in Fig. 6.2.

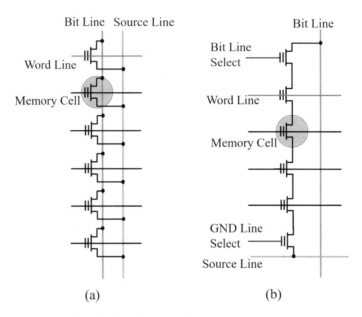

Fig. 6.2 **a** NOR and **b** NAND flash architecture [5]

6.5   Multi-level Memory Cell Concept

Storing of analog data in a floating gate memory is an old concept. It was proposed for EPROM devices in early 70s [6] and was implemented for EEPROM devices in neural networks, voice recorder etc. in early 80s. These analog devices are much more fault tolerant and can function almost properly in the presence of loss of few bits of data. But the digital devices are not as much fault tolerant and thus have limitation for a higher bit of data per cell. The concept of Multi-Level Cell (MLC) was introduced to store two bits of data per cell in digital storage devices with minimum amount of error. The MLC concept can be adopted to develop new type of flash memory that will require lower material processing cost and offer higher information density. The error associated with storing multiple bits of data in a single memory cell can be compensated by an Error Correcting Code (ECC).

6.5.1   Working Principle of Multi-valued Flash Memory

In a binary flash memory cell, which is made of a single floating gate transistor, the presence and the absence of stored or trapped charge on the floating gate provide the binary signals "0" and "1". During the programming of the device, charge or hot electrons get trapped in the floating gate of the cell. For 1 bit per cell (binary) flash devices, if the

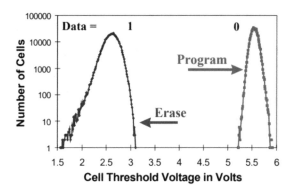

Fig. 6.3 Single bit per cell (binary) array threshold voltage histogram [7]

trapped charge is less than a certain value, the cell requires a lower threshold voltage to turn on. That is interpreted as a logic "1" stored in the cell. On the other hand, if the amount of charge is high enough, the effective threshold voltage of the device increases, and the cell requires higher gate voltage to turn on. This state is interpreted as a logic "0". The threshold voltage distributions of a half-million binary cell array block presented in [7] are illustrated in Fig. 6.3. After erasure/programming operation, the threshold voltage of each cell is calculated and the histogram of all the results are shown. For erased cells (logic 1), the threshold voltage is < 3.1 V. And for programmed cells (logic 0), threshold voltage is > 5 V.

The design of multi-level flash memory cell is based on the same concept of storing charge on the floating gate transistor. By storing/trapping different amounts of charges or hot electrons the threshold voltage of the floating gate transistor can be set at different levels, which in turn will lead to different gate voltage requirements to turn on the device. This ability to set multiple threshold voltages provides the option to achieve multiple logic levels in a single cell. The charge storing capacity of a flash memory cell is an analog property. Using a very controlled programming method, a certain amount of charge can be stored in the flash cell. If four distinct and precise levels of charge can reliably be stored in the floating gate transistor it is possible to obtain a memory cell capable of storing two bits per cell. These four distinct charge levels in the cell corresponds to four different logic levels, which can be set at 25, 50, 75 and 100% of the charge storage capacity of the cell. These four logic levels refer to four binary values, 11, 01, 10 and 00, respectively. Here, the most challenging part is the determination and reliable holding of the intermediate voltages (gaps between the charge levels) between the logic values. The read and write margins will be limited by these intermediate gaps, which will affect the ability to accurately read and write data into the cell [7]. The threshold voltage distribution of a half-million memory cell array block for 2-bit MLC is given in Fig. 6.4. This distribution is extracted from [7] for Intel StrataFlash memory. After erasure

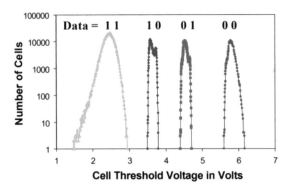

Fig. 6.4 Two bits per cell array threshold voltage histogram [7]

or programming operation to one of the three logic states, the threshold voltage for each cell is shown in the graph.

6.5.2   Planar Versus Vertical NAND

Most of the multi-level flash memories are planar in nature and are prone to scaling limitations. To overcome the scaling challenges many new approaches are being adopted by the memory industry. Among them, 3D NAND flash is the most promising one. The concept of 3D memory was successfully implemented by Samsung in the form of 3D NAND flash, where memory cells are stacked vertically. The vertical NAND flash, which has become known as the VNAND, can store three bits of data in each cell [8]. This type of Triple Level Cell (TLC) can store three bits of information and the combination of these three bits can provide eight different logic values. There are other planar and 3D TLC and Quad Level Cell (QLC) designs available in the industry [9–11]. Figure 6.5a represents the schematic diagram of a planar NAND, which is 2D structure and Fig. 6.5b represents a 3D NAND, which resembles a vertical structure [12].

Some additional designs of 3 or 4 bits per cell flash memory designs have been reported in [13–17]. The number of bits that can be contained by a cell is controlled by the different levels of charges that can reliably be stored on the floating gate. Bits per cell defines the number of bits that can be written or read from a single cell. Table 6.1 presents a comparative analysis of different MVL flash memories. Mostly MOSFET technologies are used for the implementation of these MVL flash memory cells.

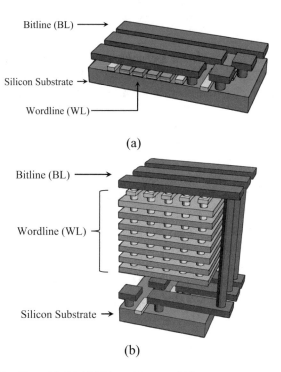

Bitline (BL) ⟶

Silicon Substrate →

Wordline (WL)

(a)

Bitline (BL) ⟶

Wordline (WL)

Silicon Substrate →

(b)

Fig. 6.5 Structure of **a** 2D and **b** 3D NAND architecture [12]

Table 6.1 Comparative analysis of different MVL flash memories

References	[13]	[14]	[15]	[16]	[17]
Bits per cell	3	3	3	3	4
Density	768 Gb	128 Gb	256 Gb	128 Gb	1 Gb
Die size	179.2 mm^2	68.9 mm^2	97.6 mm^2	146.5 mm^2	85 mm^2
Read throughput	800 Mb/s	45 μs	45 μs	90 μs	> 20 Mb
Program throughput	44 Mb/s	700 μs	660 μs	2.8 ms	3.5 Mb/s
Technology	3-D floating gate NAND flash memory	Second generation V-NAND with 32 stacked WL layers	3D NAND 48 stacked WL	NAND flash using 20 nm planar-cell technology	NROM flash

6.6   Conclusion

Continuous scaling of semiconductor technologies is leading to smaller transistor sizes and lower supply voltages. The highly scaled MOSFET devices used in flash memory must endure high operating voltages required during the "hot electron" injection between the channel and the floating gate. To lower the operating voltage the oxide layer between the control gate and the floating gate of the flash memory device must be made thinner. However, the presence of a very thin insulating layer in between the floating gate and the control gate can lead to severe unintentional disturbance of electrons and thereby resulting in erroneous programming and erasing operation. These challenges lead to increased fabrication complexity and reliability issues. The challenges and the consequences are more severe in the case of multi-level flash memories [18]. Some of the other major challenges of multi-valued flash memories are precise placement of charge during the programming, precise sensing of charge during the reading, and the storage of stable amount of charge for a long enough time [7].

In addition to the conventional MOSFET based flash memory designs, other emerging technologies like FinFET and CNTFET are being explored to implement floating gate devices for flash memory [19–21].

References

1. https://www.techopedia.com/definition/24481/flash-memory.
2. "1971: Reusable semiconductor ROM introduced". Computer History Museum. Retrieved 25 December 2020 (https://www.computerhistory.org/storageengine/reusable-semiconductor-rom-introduced/#:~:text=1971%20Intel%20announced%20the%201702,pattern%20with%20ultra%2Dviolet%20light).
3. Masuoka F, Iizuka H, inventors; Toshiba Corp, assignee. Semiconductor memory device and method for manufacturing the same. United States patent US 4,531,203. 1985 Jul 23.
4. Student circuit. Retrieved on 25th December 2020 (https://www.student-circuit.com/learning/year3/embedded-systems/what-is-the-difference-between-nand-and-nor-flash-memory/).
5. Embedded.com. Retrieved on 25th December 2020 (https://www.embedded.com/flash-101-nand-flash-vs-nor-flash/).
6. Frohman-Bentchowsky, Floating Gate Solid State Storage Device and Methodology for Charging and Discharging Same, U.S. Patent #3,755,721, Aug. 28, 1973.
7. Atwood G, Fazio A, Mills D, Reaves B. Intel StrataFlash memory technology overview. Intel Technology Journal. 1997;4:1–8.
8. Samsung.com. Retrieved on 25th December 2020 (https://www.samsung.com/us/aboutsamsung/company/history/).
9. Slashgear.com, "SanDisk ships world's first memory cards with 64 gigabit X4 NAND flash", Shane McGlaun on Oct 13, 2009. Retrieved on 25th December 2020 (https://www.slashgear.com/sandisk-ships-worlds-first-memory-cards-with-64-gigabit-x4-nand-flash-1360217/).

10. Anandtech.com, "ADATA Reveals Ultimate SU630 SSD: 3D QLC for SATA", Anton Shilov on November 15, 2018. Retrieved on 25th December 2020 (https://www.anandtech.com/show/13606/adata-ultimate-su630-ssd-3d-qlc-for-sata).

11. Anandtech.com, "The Intel SSD 660p SSD Review: QLC NAND Arrives For Consumer SSDs", Billy Tallis on August 7, 2018 (https://www.anandtech.com/show/13078/the-intel-ssd-660p-ssd-review-qlc-nand-arrives).

12. nvm durance, "The 2D NAND problems NVMdurance solves", Mark Lapedus on March 13th, 2017, Retrieved on 1st January, 2022 (https://www.nvmdurance.com/2d-nand-problems-nvm durance-solves/).

13. Tanaka T, Helm M, Vali T, Ghodsi R, Kawai K, Park JK, Yamada S, Pan F, Einaga Y, Ghalam A, Tanzawa T. 7.7 A 768Gb 3b/cell 3D-floating-gate NAND flash memory. In 2016 IEEE International Solid-State Circuits Conference (ISSCC) 2016 Jan 31 (pp. 142–144). IEEE.

14. Jeong W, Im JW, Kim DH, Nam SW, Shim DK, Choi MH, Yoon HJ, Kim DH, Kim YS, Park HW, Kwak DH. A 128 Gb 3b/cell V-NAND flash memory with 1 Gb/s I/O rate. IEEE Journal of Solid-State Circuits. 2015 Sep 21;51(1):204–12.

15. Kang D, Jeong W, Kim C, Kim DH, Cho YS, Kang KT, Ryu J, Kang KM, Lee S, Kim W, Lee H. 256 Gb 3 b/cell V-NAND flash memory with 48 stacked WL layers. IEEE Journal of Solid-State Circuits. 2016 Oct 24;52(1):210–7.

16. Naso G, Botticchio L, Castelli M, Cerafogli C, Cichocki M, Conenna P, D'Alessandro A, De Santis L, Di Cicco D, Di Francesco W, Gallese ML. A 128Gb 3b/cell NAND flash design using 20nm planar-cell technology. In 2013 IEEE International Solid-State Circuits Conference Digest of Technical Papers 2013 Feb 17 (pp. 218–219). IEEE.

17. Polansky Y, Lavan A, Sahar R, Dadashev O, Betser Y, Cohen G, Maayan E, Eitan B, Ni FL, Ku YH, Lu CY. A 4b/cell NROM 1Gb data-storage memory. In 2006 IEEE International Solid State Circuits Conference-Digest of Technical Papers 2006 Feb 6 (pp. 448–458). IEEE.

18. Zeraatkar N. Flash Memory Technology. September 2006, 9th Iranian Student Conference on Electrical Engineering, Tehran (https://civilica.com/doc/48624/).

19. Zhu, Huilong. "FinFET flash memory device with an extended floating back gate." U.S. Patent No. 7,619,276. 17 Nov. 2009.

20. Seike, Kohei, et al. "Floating-gated memory based on carbon nanotube field-effect transistors with Si floating dots." Japanese Journal of Applied Physics 53.4S (2014): 04EN07.

21. Lee, Sang Wook, et al. "A fast and low-power microelectromechanical system-based non-volatile memory device." Nature Communications 2.1 (2011): 1–6.

Ternary Content Addressable Memory

7

7.1 Introduction

Content-addressable memory (CAM) is a special type of memory used in very-high-speed search applications. The look-up table function implementation for a CAM in silicon is conceptually similar to the implementation of array logic in data structures. However, the output in CAM is highly simplified. The input search data enters the CAM, compares against a stored data table, and returns the matching data address. Once a key is passed to a CAM sub-system, it returns the corresponding value of that key. In CAM, a single clock cycle is needed to look up an entry, whereas it requires multiple clock cycles in RAM. This is the most important feature of CAM, making it faster than other hardware- and software-based search systems [1].

CAM is mostly used in networking devices where information forwarding, and routing table operations require faster procedures. Cache memory also uses associative memory like CAM, where both address and content are stored side by side. If the target address matches, cache memory gives out the content of that address. CAMs can be used in many other applications where high search speed is required. Some of example applications are Huffman coding/decoding [2, 3], parametric curve extraction [4], Lempel–Ziv compression [5], Hough transformation [6], and image coding [7]. CAMs are commercially used in network routers to classify Internet Protocol (IP) packets and forward it to the next equipment [8, 9]. In network communication (like internet) channels, each message like an e-mail or a web page is first broken into smaller data packets of several hundred bytes. It transmits each data packet individually through the network. These packets are sent in the middle of the network (called routers) from the source through the nodes. In the end, towards the destination, packets are reconnected to reproduce the original message. The router's job is to compare the destination address and choose the appropriate ones among all possible routes. CAMs are a good alternative for implementing this lookup operation

Z. T. Sandhie et al., *Beyond Binary Memory Circuits*, Synthesis Lectures
on Digital Circuits & Systems, https://doi.org/10.1007/978-3-031-16195-7_7

because of their ability to search quickly. However, higher speed in CAM comes at the cost of two critical design overheads—large silicon area and higher power consumption. The demand for larger CAM sizes in some emerging application devices exacerbates the power issue. Reducing power consumption in high-capacity CAMs without sacrificing speed or field, is a critical research problem to focus on.

7.2   Operation of Content-Addressable Memory (CAM)

Here, a brief description of the design and operation of CAM as a data packet forwarding device is provided. Figure 7.1 [1] shows a simplified block diagram of a CAM, where the shaded box indicates the search word which matches the location (w-2). The row match results are provided by the matchlines. The encoder output is the encoded version of the latch location using log_2w bits.

Input into the system is the search word transmitted to the searchline in the table of stored information. While the number of bits in a CAM word is usually large, existing implementations range from 36 to 144 bits. A typical CAM takes the size of a table from a few hundred entries to 32K entries, corresponding to 7 bits to 15 bits address sizes. Each stored word has a match, which indicates that the search term and the stored word are identical (match case) or different (no match or miss). The machines are fed to an encoder that creates a binary match position with the matchline in the match state. System uses an encoder where only a single match can be expected. A priority encoder is used instead of the usual encoder in CAM applications where more than one word

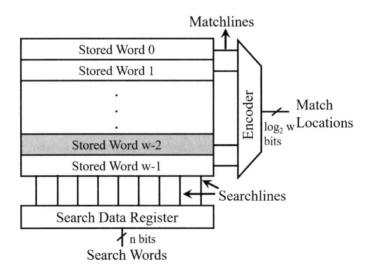

Fig. 7.1 Conceptual diagram of a content-addressable memory (CAM) containing w words [1]

can match. A priority encoder selects the highest priority match location to map match outcomes. Usually, the highest priority is given to words in lower address locations. A hit signal is flagged in case there is no match location in the CAM space. The overall function of CAM is to take a search term and return the matched memory information. It can be portrayed as a fully programmable arbitrary mapping operation of a large space of input search terms to match smaller output locations. In terms of operation, CAM is the opposite of Random-Access-Memory (RAM). RAM looks up for data using the memory address and then returns the data from the address. CAM does the opposite of lookups. A function calls CAM by passing a key containing the data word structure and gives the CAM lookup memory addresses.

In traditional ASIC hardware, a CAM cell usually contains two SRAM cells. The SRAM based implementation of CAM leads to the usage of wide silicon gates, which causes high power consumption for fast switching. Higher power consumption leads to excessive heat that limits integration density and CAM sizes. This is the main bottleneck that is preventing the implementation of larger CAM [10] in state-of-the-art applications.

7.3   Binary Versus Ternary CAM

A Ternary Content-Addressable Memory (TCAM) is a specific type of high-speed memory which can search its' whole contents in a single clock cycle. In binary CAM, the stored or searched data is comprised of only "0" or "1". On the other hand, ternary CAM can store and search data with three different outputs, 0, 1, and X (don't care). The advantage of the "X" or don't care state is that it enables the ternary CAM to execute a wide-ranging search system based on pattern matching, whereas binary CAM does exact matches using only "0" and "1" [11].

Figure 7.2 shows the schematic diagram of a conventional binary CAM [12]. It consists of a cross-coupled inverter pair. The data is written in the memory cell through the access transistors Q_1 and Q_2. During the search phase, the matchline (ML) is pre-charged to V_{DD} before being evaluated. The evaluation phase depends on the arrangement of the transistors Q_3 to Q_6. When the stored data at node D is a match with the searchline (SL) data, no direct path exits for the ML to discharge to the ground. This causes the ML to remain at a high voltage level which indicates a match. On the other hand, if the stored data doesn't match with the data at SL, ML discharges to a low voltage. A similar configuration of Fig. 7.2 can be used to implement a ternary CAM. The major difference is that the data is stored in two separate cells which results in a 16-transistor design. The above-mentioned design is a static storage-based design which contains several advantages like higher noise margin, improved data retention, etc.

A Ternary CAM (TCAM) can also be implemented as dynamic memory. The main advantage of a dynamic memory cell is the significantly reduced number of transistors. Figure 7.3 shows a dynamic TCAM cell containing 4 transistors [13]. In the figure, Q_1

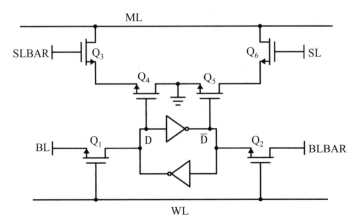

Fig. 7.2 Schematic diagram of a conventional binary static CAM [12]

and Q_2 are two access transistors, which are gated by a "Write" signal. When "Write" $=$ V_{DD} the data from the bitlines (BL and BLBAR) are passed to the storage nodes D_1 and D_0 respectively. Two dynamic nodes D_1 and D_2 are generated from the gate capacitance of the transistors Q_3 and Q_4. Storing (0, 1) in (D_1, D_0) is interpreted as logic 0 stored in the memory cell. Storing (1, 0) in (D_1, D_0) represents logic 1 stored in the device. And storing (0, 0) is considered as the don't care stage. During the search operation, the value on the "Write" signal is set to ground. Then a comparison of data is done between the bitline (BL) and the storage node. Like the conventional binary CAM, the ML is precharged to V_{DD} and then evaluated. The ML stays at high voltage during a match and discharges to low voltage (ground) during a mismatch.

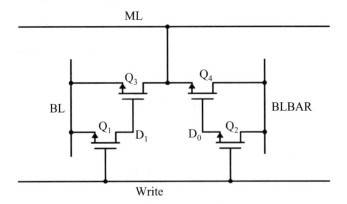

Fig. 7.3 Schematic diagram of a 4-transistor dynamic TCAM cell [13]

7.4   Analysis

As mentioned earlier, CAM's operation is the opposite of a RAM. To search for a data from a RAM the operating system must provide the address of the memory where the data is stored. On the other hand, to access a stored data on CAM, the content of the CAM itself is queried, upon which the memory retrieves the address where the data is stored. In terms of operation speed, CAM (and TCAM) are considerably faster than the RAM. However, due to the larger cell size, CAM is expensive, consume higher power, and dissipates more heat [14]. Several different design approaches have been proposed to implementation TCAM. The initial designs were based on MOSFET and MOSFET-resistor. In recent time, new TCAM designs based on MOSFET-memristor, CNTFET, and Metal-Tunnel-Junction (MTJ) have been proposed.

TCAM cell offers many benefits compared to the binary CAM cell. However, there are some crucial challenges regarding the fabrication of TCAM cells with large storage capacity. The most critical issue is the requirement of larger gate size and chip area to implement high capacity TCAM. A conventional TCAM cell needs 16 transistors [1]. Though different novel approaches have been proposed to minimize the transistor count in TCAM cell, still there is a stability issue if the cell is being used in large-scale chips. Some examples of TCAM cell designs with lower number of transistors count are 12T SRAM based design [15], 6T-2C DRAM based design [16], etc. Due to the volatile nature of these TCAM designs, they consume high standby power [17]. Another critical issue for the TCAM cell designs is higher power dissipation. Since the search operation in a TCAM cell is fully parallel, it needs a significant amount of power for the search to be completed. This large power dissipation lowers the reliability of the chip and increases the packaging cost [16]. Different techniques to reduce the power dissipation in TCAM cell have been proposed [1]. However, the resulting designs suffer from either large area overhead or inadequate noise immunity. Another problem with the TCAM circuit implementation is low yield [16].

For successful design and fabrication of TCAM chips the above-mentioned issues must be addressed. One way to address some of these challenges is to design the TCAM cell as non-volatile memory. Recently, some non-volatile designs have been proposed using MTJ [17–20], resistive storage device like memristor [11, 21], and other emerging technologies. In an MTJ-based TCAM cell, the performance depends significantly on the functionality of MTJ. Special care should be taken so that the data stored in the TCAM is not changed while the evaluating a match or mismatch stage. The magnetization of a MTJ can flip in the presence of a small amount of current flow during a long period of time [22]. To increase the reliability of MTJ, clock-gating is introduced in [23], so that the stored data in MTJ is not hampered even if the search input is unchanged for a long time. Due to this clock-gating scheme, the search inputs are kept at zero voltage during the pre-charge stage of ML, which prevents the damage of the stored data. Additionally, the MTJ reliability can be increased by using a material with higher thermal stability factor. The cell area is not affected by the MTJ as the MTJ can be stacked over the MOS transistors [18]. A comparative analysis of different TCAM cell designs is provided in Table 7.1.

The comparative table discusses the few basic features of the prevailing TCAM designs. Here, we can see that the design approach of a TCAM varies from using only transistor or a combination of transistor with capacitor or resistor or MTJ to obtain the required result. Most of the prevailing design uses MOS type transistor in their design. The transistor-MTJ based designs are mostly non-volatile in nature. Based on the design structure and number of transistor or MTJ, other qualities like cell area, matching/searching capability etc. varies.

Table 7.1 Comparative analysis of different TCAM design

References	[15]	[16]	[21]	[11]	[17]	[18]	[19]	[20]
Cell feature	12 T	6T-2C	5T-2R (5 transistor-2 memristor)	2T-2R (2 transistor-2 resistive storage)	4T-2MTJ	6T-2MTJ	10T-4MTJ	12T-2MTJ
Technology	0.18 μm SRAM based volatile design	0.13 μm 6Cu eDRAM volatile process	–	90 nm CMOS/PCM nonvolatile TCAM	90 nm CMOS-100 nm MTJ nonvolatile TCAM	90 nm 1P5M CMOS/MTJ nonvolatile TCAM	45 nm CMOS/MTJ nonvolatile TCAM	65 nm CMOS/MTJ nonvolatile TCAM
Supply voltage (V)	1.2	Local searching: 1 V, peripheral circuits: 1.5 V, IO circuits: 2.5 V	1 (search) 2.5 (write)	0.75–1.2 (search) 2.5 (write)	1.2	1.2	1	1.2
Cell area (μm^2)	17.54	3.59	–	0.41	3.14	10.35	2.78	3.65
Array configuration	–	–	–	–	32 bits × 64 words	32 bits × 64 words	–	–
T_{search}/match delay (ns)	3	–	–	1.9@1.2 V	2.5	0.29	1	0.6
E_{search}/bit (fJ)	–	–	–	–	–	1.04	40.5	4.7
E_{write}/bit (pJ)	–	–	–	–	–	–	0.87	0.45

References

1. Pagiamtzis K, Sheikholeslami A. Content-addressable memory (CAM) circuits and architectures: A tutorial and survey. IEEE journal of solid-state circuits. 2006 Feb 27;41(3):712–27.
2. E. Komoto, T. Homma, and T. Nakamura, "A high-speed and compactsize JPEG Huffman decoder using CAM," in Symp. VLSI Circuits Dig. Tech. Papers, 1993, pp. 37–38.
3. L.-Y. Liu, J.-F. Wang, R.-J. Wang, and J.-Y. Lee, "CAM-based VLSI architectures for dynamic Huffman coding," IEEE Trans. Consumer Electron., vol. 40, no. 3, pp. 282–289, Aug. 1994.
4. M. Meribout, T. Ogura, and M. Nakanishi, "On using the CAM concept for parametric curve extraction," IEEE Trans. Image Process., vol. 9, no. 12, pp. 2126–2130, Dec. 2000.
5. B. W. Wei, R. Tarver, J.-S. Kim, and K. Ng, "A single chip Lempel-Ziv data compressor," in Proc. IEEE Int. Symp. Circuits Syst. (ISCAS), vol. 3, 1993, pp. 1953–1955.
6. M. Nakanishi and T. Ogura, "Real-time CAM-based Hough transform and its performance evaluation," Machine Vision Appl., vol. 12, no. 2, pp. 59–68, Aug. 2000.
7. S. Panchanathan and M. Goldberg, "A content-addressable memory architecture for image coding using vector quantization," IEEE Trans. Signal Process., vol. 39, no. 9, pp. 2066–2078, Sep. 1991.
8. T.-B. Pei and C. Zukowski, "VLSI implementation of routing tables: tries and CAMs," in Proc. IEEE INFOCOM, vol. 2, 1991, pp. 515–524.
9. N.-F. Huang, W.-E. Chen, J.-Y. Luo, and J.-M. Chen, "Design of multi-field IPv6 packet classifiers using ternary CAMs," in Proc. IEEE GLOBECOM, vol. 3, 2001, pp. 1877–1881.
10. EtherealMind.com, "Basics: What is Content Addressable Memory (CAM)?" by Greg Ferro on 5th July, 2016, Retrieved on 25th Dec, 2020 (https://etherealmind.com/basics-what-is-content-addressable-memory-cam/).
11. Li, Jing, et al. "1 Mb 0.41 μm^2 2T-2R cell nonvolatile TCAM with two-bit encoding and clocked self-referenced sensing." IEEE Journal of Solid-State Circuits 49.4 (2013): 896–907.
12. Hellkamp D, Nepal K. Metallic tube-tolerant ternary dynamic content-addressable memory based on carbon nanotube transistors. Micro & Nano Letters. 2015 Apr 23;10(4):209–12.
13. Delgado-Frias JG, Yu A, Nyathi J. A dynamic content addressable memory using a 4-transistor cell. In Proceedings of the Third International Workshop on Design of Mixed-Mode Integrated Circuits and Applications (Cat. No. 99EX303) 1999 Jul 28 (pp. 110–113). IEEE.
14. "Ternary content-addressable memory (TCAM)" by Jessica Scarpati, Retrieved on 25th Dec, 2020 (https://searchnetworking.techtarget.com/definition/TCAM-ternary-content-addressable-memory).
15. Arsovski, Igor, Trevis Chandler, and Ali Sheikholeslami. "A ternary content-addressable memory (TCAM) based on 4T static storage and including a current-race sensing scheme." IEEE Journal of Solid-State Circuits 38.1 (2003): 155–158.
16. Noda, Hideyuki, et al. "A cost-efficient high-performance dynamic TCAM with pipelined hierarchical searching and shift redundancy architecture." IEEE Journal of Solid-State Circuits 40.1 (2005): 245–253.
17. Matsunaga, Shoun, et al. "A 3.14 um 2 4T-2MTJ-cell fully parallel TCAM based on nonvolatile logic-in-memory architecture." 2012 Symposium on VLSI Circuits (VLSIC). IEEE, 2012.
18. Matsunaga, Shoun, et al. "Fully parallel 6T-2MTJ nonvolatile TCAM with single-transistor-based self match-line discharge control." 2011 Symposium on VLSI Circuits-Digest of Technical Papers. IEEE, 2011.
19. Song, Byungkyu, et al. "A 10T-4MTJ nonvolatile ternary CAM cell for reliable search operation and a compact area." IEEE Transactions on Circuits and Systems II: Express Briefs 64.6 (2016): 700–704.

20. Cho, Dooho, Kyungmin Kim, and Changsik Yoo. "A non-volatile ternary content-addressable memory cell for low-power and variation-toleration operation." IEEE transactions on magnetics 54.2 (2017): 1–3.
21. Zheng, Le, Sangho Shin, and Sung-Mo Steve Kang. "Memristors-based ternary content addressable memory (mTCAM)." 2014 IEEE International Symposium on Circuits and Systems (ISCAS). IEEE, 2014.
22. Y. Zhang et al., "Compact modeling of perpendicular-anisotropy CoFeB/MgO magnetic tunnel junctions," IEEE Trans. Electron Devices, vol. 59, no. 3, pp. 819–826, Mar. 2011.
23. Gupta MK, Hasan M. Robust high speed ternary magnetic content addressable memory. IEEE Transactions on Electron Devices. 2015 Mar 10;62(4):1163–9.

Printed in the United States
by Baker & Taylor Publisher Services